Path Dependence and New Path Creation in Renewable Energy Technologies

Why are old technologies persisted with after better alternatives have been invented? This book examines this question, a central concern of evolutionary economics, specifically focusing on renewable energy technologies. The concept of path dependence is used to analyse why and how technological development can become locked-in to inefficient ways of doing things. This book shows how lock-in can be avoided by the creation of new technological pathways.

The chapters focus on the comparatively recent introduction of new wind turbine technologies for the generation of carbon free electricity. This case study provides valuable lessons in understanding the issues confronting inventors attempting to commercialise their new ideas in the form of innovations in the face of historically established conventional technologies. It is also set within the critical debate on climate change and the need to de-carbonise energy supplies in order to stop further man-made deterioration in the global environment.

This book was originally published as a special issue of *European Planning Studies.*

James Simmie is Professor of Innovation Studies at Oxford Brookes University, UK. His work is focused on the relationships between innovation, productivity and the competitiveness of urban regions. It is developed within the theoretical framework of evolutionary economics and geography. Over the years he has published widely on these themes. Some of his more recent publications are available to download from http://planning.brookes.ac.uk/staff/jamessimmie.html

Path Dependence and New Path Creation in Renewable Energy Technologies

Edited by
James Simmie

Routledge
Taylor & Francis Group
LONDON AND NEW YORK

First published in paperback 2024

First published 2014
by Routledge
4 Park Square, Milton Park, Abingdon, Oxon, OX14 4RN

and by Routledge
605 Third Avenue, New York, NY 10158

Routledge is an imprint of the Taylor & Francis Group, an informa business

Publisher's Note
The publisher accepts responsibility for any inconsistencies that may have arisen during the conversion of this book from journal articles to book chapters, namely the possible inclusion of journal terminology.

Disclaimer
Every effort has been made to contact copyright holders for their permission to reprint material in this book. The publishers would be grateful to hear from any copyright holder who is not here acknowledged and will undertake to rectify any errors or omissions in future editions of this book.

British Library Cataloguing in Publication Data
A catalogue record for this book is available from the British Library

ISBN: 978-1-138-02417-5 (hbk)
ISBN: 978-1-03-292888-3 (pbk)
ISBN: 978-1-315-77590-6 (ebk)

DOI: 10.4324/9781315775906

Typeset in Times New Roman
by Taylor & Francis Books

Contents

Citation Information vii

1. Introduction: Path Dependence and New Path Creation in Renewable Energy Technologies
James Simmie 1

Part I: Theories

2. Path Creation: Co-creation of Heterogeneous Resources in the Emergence of the Danish Wind Turbine Cluster
Peter Karnøe & Raghu Garud 5

3. Path Dependence and New Technological Path Creation in the Danish Wind Power Industry
James Simmie 25

4. Comparing Alternative Path Creation Frameworks in the Context of Emerging Biofuel Fields in the Netherlands, Sweden and Finland
Raimo Lovio & Paula Kivimaa 45

Part II: Geographies

5. Renewable Energy Technology and Path Creation: A Multi-scalar Approach to Energy Transition in the UK
Jürgen Essletzbichler 63

6. Transversality and Transition: Green Innovation and New Regional Path Creation
Philip Cooke 89

7. From the Old Path of Shipbuilding onto the New Path of Offshore Wind Energy? The Case of Northern Germany
Dirk Fornahl, Robert Hassink, Claudia Klaerding, Ivo Mossig & Heike Schröder 107

8. Spatial Processes of Industry Emergence: US Wind Turbine Manufacturing Industry
Gregory Theyel 129

Part III: Policy

9. Clean Energy Technology and the Role of Non-Carbon Price-Based Policy: An Evolutionary Economics Perspective
Nicholas Howarth 143

Index 165

Citation Information

The chapters in this book were originally published in *European Planning Studies*, volume 20, issue 5 (May 2012). When citing this material, please use the original page numbering for each article, as follows:

Chapter 1
Introduction: Path Dependence and New Path Creation in Renewable Energy Technologies
James Simmie
European Planning Studies, volume 20, issue 5 (May 2012) pp. 729–732

Chapter 2
Path Creation: Co-creation of Heterogeneous Resources in the Emergence of the Danish Wind Turbine Cluster
Peter Karnøe & Raghu Garud
European Planning Studies, volume 20, issue 5 (May 2012) pp. 733–752

Chapter 3
Path Dependence and New Technological Path Creation in the Danish Wind Power Industry
James Simmie
European Planning Studies, volume 20, issue 5 (May 2012) pp. 753–772

Chapter 4
Comparing Alternative Path Creation Frameworks in the Context of Emerging Biofuel Fields in the Netherlands, Sweden and Finland
Raimo Lovio & Paula Kivimaa.
European Planning Studies, volume 20, issue 5 (May 2012) pp. 773–790

Chapter 5
Renewable Energy Technology and Path Creation: A Multi-scalar Approach to Energy Transition in the UK
Jürgen Essletzbichler
European Planning Studies, volume 20, issue 5 (May 2012) pp. 791–816

Chapter 6
Transversality and Transition: Green Innovation and New Regional Path Creation
Philip Cooke
European Planning Studies, volume 20, issue 5 (May 2012) pp. 817–834

Chapter 7
From the Old Path of Shipbuilding onto the New Path of Offshore Wind Energy? The Case of Northern Germany
Dirk Fornahl, Robert Hassink, Claudia Klaerding, Ivo Mossig & Heike Schröder
European Planning Studies, volume 20, issue 5 (May 2012) pp. 835–856

Chapter 8
Spatial Processes of Industry Emergence: US Wind Turbine Manufacturing Industry
Gregory Theyel
European Planning Studies, volume 20, issue 5 (May 2012) pp. 857–870

Chapter 9
Clean Energy Technology and the Role of Non-Carbon Price-Based Policy: An Evolutionary Economics Perspective
Nicholas Howarth
European Planning Studies, volume 20, issue 5 (May 2012) pp. 871–892

Please direct any queries you may have about the citations to clsuk.permissions@cengage.com

INTRODUCTION

Path Dependence and New Path Creation in Renewable Energy Technologies

JAMES SIMMIE

Department of Planning, Oxford Brookes University, Headington Campus, Gipsy Lane, Oxford OX3 0BP, UK

The papers in this special edition of European Planning Studies present research on the theoretical, spatial and policy issues surrounding the need to create new technological pathways in order to mitigate the effects of climate change. They start from the assumption that the scientific case for climate change and the contribution of the human race to that phenomenon has been made. The failure of the Copenhagen Conference on Climate Change in 2009, however, demonstrated the immense difficulties associated with moving from the scientific evidence to effective action to slow down and reverse the contemporary rates of climate change.

The theoretical starting point for the papers in this special edition is that contemporary forms of economic activity that contribute to global warming are often locked-in to historical and path-dependent technological development trajectories. It is argued that this is one significant reason why it is proving so difficult to change rapidly from, for example, fossil fuel-based energy systems to the general use of renewable energy technologies (RETs). From this perspective the key question addressed by the papers presented here is "given the historical path dependent economic development of fossil fuel based energy systems, how can new pathways be created in renewable energy technologies"? This question is addressed from three different perspectives.

First, a group of three papers analyse the theoretical possibilities for new path creation. In the first of these Peter Karnøe and Raghu Garud argue that in order to theorize new path creation we need an ontological shift from the economic approach adopted in canonical path dependence theory to a sociological perspective. This is because, in the former, the movement towards lock-in is driven by disembodied economic forces such as contingency and self-reinforcement that provide no explanation of agency. In contrast, Karnøe and Garud argue that initial conditions and new path creation are constructed by reflexive agents such as entrepreneurs. Therefore, we need sociological analyses that explain how such agents go about constructing the perceived realities and conditions into which they introduce new technologies. This approach takes as its starting point sociological studies of science and technology. They illustrate their arguments using an historical and sociological analysis of the emergence of the Danish wind power industry.

In the second paper, James Simmie takes up these arguments and suggests that it is unlikely that the complexities of new path creation can be explained adequately from one ontological perspective. A key requirement of any such explanation is that it explains how the processes of invention and innovation can create new technological pathways. This requires a combination of economic and sociological explanations of how inventors, innovators and diffusion agents overcome contemporary path-dependent economic barriers, technological paradigms, institutional hysteresis and the power of vested interests in existing technological regimes in order to introduce and diffuse the use of a new technology. Simmie argues that frequently the introduction of a new technology takes place in a niche that shelters it from the full force of the existing path-dependent systems. This is also illustrated empirically by a socio-economic historical analysis of the early emergence and subsequent evolution of the Danish wind power industry.

The third paper in this group also takes up the theme of the significance of the deployment of different ontologies to explain how innovations and new technological paths are created. Raimo Lovio and Paula Kivimaa compare two different theoretical approaches to understanding the processes of new path creation. These are strategic niche management and its associated multi level perspective (MLP), and the technological innovation systems approach. They hypothesize that the empirical results obtained by employing these two theories are more similar than the original theorizations would suggest. They illustrate these arguments by an analysis of the development of bio fuels in Finland, the Netherlands and Sweden.

A second group of papers seek to explain the uneven spatial development of new technological pathways. The first of these papers by Jurgen Essletzbichler makes the case for adopting a geographical approach to energy transition theory and policy. He takes as his starting point the MLP on energy transitions. It is argued that niches, technological regimes and economic landscapes play out differently and have different characteristics in different regions. From this perspective it is important to analyse the regional and local geographies of energy transitions. These arguments are illustrated by comparing the uneven regional development of RETs in Denmark, Germany, Spain and the UK. The first three lead the transition to new forms of energy in Europe. As ever, a few UK regions have high rates of invention but lag behind their European counterparts in turning them in to working innovations.

Phil Cooke takes up the argument that the creation of new economic pathways in RETs takes place in some types of region and not in others. One of the possible reasons for this is that regions are differentiated by the different types of knowledge found in them. This arises from previous rounds of economic and industrial development. In some cases the existing knowledge in a given region can be reconfigured to branch out of previous path-dependent trajectories. For this to happen there needs to be some element of relatedness between the varieties of old knowledge present in a region. Cooke refers to this transversality. He illustrates the process by examining the development of RETs in North Jutland, Denmark, and Norrland in Sweden. In both cases new pathways in several RETs have been created out of their previous path-dependent industrial and knowledge bases.

In the literature there is some argument concerning the general significance of related variety and branching out of existing economic pathways in order to create new ones. Dirk Fornahl, Robert Hassink, Claudia Klaerding, Ivo Mossig, and Heike Schröder address this debate. They analyse whether the new industrial path of offshore wind energy emerged out of existing paths, mainly shipbuilding, in the five states of coastal Germany, namely Bremen, Lower Saxony, Hamburg, Schleswig-Holstein and Mecklenburg-Vorpommern. They conclude that shipbuilding only indirectly affected the emergence of the new industrial development path of offshore the wind energy industry in northern Germany. Instead, the new pathway in offshore wind power was created in this area mainly by existing onshore wind firms or by other industries that have diversified in to offshore wind energy.

In the final geographic analysis of new path creation, Gregory Theyel analyses the location patterns of the US wind turbine manufacturing industry at the level of the American states. He points to the importance of the locational resources in different states in the creation of new forms of production. Local demand conditions are shown to be important. In contrast to many of the European examples, government policies are shown to have had only a weak influence on the emergence of the new industry.

In the third part of this special edition we turn to the effects of government policies on generating new economic pathways in RETs. Eric Knight and Nicholas Howard focus on the role of carbon pricing in developing a market-led response to low carbon energy innovation. Taking an evolutionary economics perspective they make a general case as to why price mechanisms alone are insufficient to support new energy technologies coming to market. They argue that the primary reason for this is that RETs face unique investment barriers in displacing existing fossil fuel-based energy systems. For guidance on possible approaches to non-carbon price-based policies that seek to tackle these barriers they analyse case studies from Asia which is a region that has experienced a strong uptake in climate policy in recent years.

The papers brought together in this edition of European Planning Studies were initially presented at a research workshop on New Path Creation held at Trinity College Oxford in September 2010. The workshop was funded by a grant from the Oxford Brookes University Central Research Fund. I should like to thank Dr Juliet Carpenter for all her help in organizing the workshop. I should also like to thank all the participants who have substantially revised their papers for inclusion in this special edition together with all the referees who provided comments for their friendly and stimulating collaboration in this project.

Path Creation: Co-creation of Heterogeneous Resources in the Emergence of the Danish Wind Turbine Cluster

PETER KARNØE* & RAGHU GARUD**

*Copenhagen Business School, Denmark, **Penn State University, USA

ABSTRACT *This paper employs path creation as a lens to follow the emergence of the Danish wind turbine cluster. Supplier competencies, regulations, user preferences and a market for wind power did not pre-exist; all had to emerge in a tranformative manner involving multiple actors and artefacts. Competencies emerged through processes and mechanisms such as co-creation that implicated multiple learning processes. The process was not an orderly linear one as emergent contingencies influenced the learning processes. An implication is that public policy to catalyse clusters cannot be based on an assumption that linear learning dynamics will unfold.*

The formation of the Danish wind turbine cluster has been the subject of much research because of its global growth and success (Jacobsson & Bergek, 2003; Garud & Karnøe, 2003; Kemp *et al.*, 2004; Karnøe, 1996, 1999a, 1999b). Since the early 1980s, Danish wind turbines have commanded a high market share in the global wind turbine industry. Moreover, companies from around the world have located R&D centres in the Danish wind power hub. Recently, in 2006, Siemens bought a leading Danish wind turbine company to build a strong position for wind power in their energy business.

Many argue that state interventions created the wind power market as state subsidies were offered to idealistic users who were motivated to use wind power for reasons ranging from their aversion to nuclear energy to their desire to reduce their dependence on imported fuel. State intervention also encouraged research and experimentation across geographically co-located entrepreneurs and it was also involved in the setting up of an approval of wind turbine designs by a test and research station. In sum, state intervention stimulated both demand pull and technology push.

While there are merits to these arguments, they nevertheless miss out on critical micro-dynamics that have to do with learning and agency. The regional setting for the wind turbine cluster was an industrial district composed of diverse SMEs with a culture for hands-on engineering in machine design. Equally importantly, there was present a culture for collaboration and interactive learning between producers, suppliers, national regulators and evaluators (Garud & Karnoe, 2003; Lundvall, 1992).

The Danish wind turbine cluster originated when new techno-economic projects such as wind turbines became associated with the emerging "alternative energy discourse". While departing from existing tools, skills and components, entrepreneurs confronted several challenges in their efforts at mobilizing support for funding and for developing their competencies. It was not easy to activate and enrol suppliers to embark on exploratory learning processes. Moreover, a political "plan" for R&D support and market subsidies for small-scale wind turbines was not part of the first official energy plan. All the while, "translation" (Callon, 1991) was involved to transform disinterested and hostile forces into supportive ones. For example, translation was involved when: (a) the electrical utilities agreed to grid-connect and pay for wind power, (b) the energy policy shifted from coal and nuclear power to include wind power and support small-scale wind turbines with a test and research centre (TRC) and market subsidies; (c) the natural landscape was categorized shaping the location of wind turbines and (c) the "Wind Atlas" became a calculative device that transformed the wind speed in different landscape types into an estimated energy production.

To capture the learning processes and the agencies involved in redirecting business activities and resources, we offer the concept of path creation (Garud & Karnøe, 2001; Garud *et al.*, 2010). This perspective breaks ontologically from the notion of agency in the path dependence literature (e.g. David, 1993) by considering translational processes involved in any journey, including cluster formation. It embraces a relational ontology that does not consider structures as being constraining, but instead one that provides the very fabric for transformation and change.

Path Creation and Clusters

A cluster is defined as a "concentration of inter-dependent firms within the same or adjacent industrial sectors in a small geographical area" (Isaksen & Hauge, 2002; Porter, 2000). Over time, research on cluster emergence has moved away from a pure "agglomeration model" (Silicon Valley) to an "industrial complex model" (Silicon Glenn, Scottish Electronics Industry) and now more recently to a "social network model" (Italian industrial districts) (Iammarino & McCann, 2006, p. 1029). Most cluster research takes into account tacit and explicit knowledge, technological regimes, industrial structure, as well as social networks in understanding the dynamics of cluster formation as emerging ecologies of action and knowledge (Brown & Duguid, 2000, p. 164).

While there are merits to these models, most embrace a "path dependence" (David, 1993) view that overlooks the agencies involved in bringing about new clusters in industrial regions. In this vein, Martin (2010) has argued that "lock-in" associated with canonical forms of path dependence is not adequate to understand the ongoing transformations of regions. Those who have used path dependence view organizational routines, skills and competencies, regulations and social norms mainly as constraining action. Such a structural view of agencies is evident in cluster studies where "growth opportunities are assumed to be shaped and constrained by path dependence, i.e. by the inheritance of local structural characteristics from past knowledge accumulation and learning, and there may often be geographically determined" (Iammarino & McCann, 2006, p. 1027).

Path creation differs from this view by flattening the distinction between agency and structure and by entertaining a relational perspective where agency emerges from interaction in sociotechncial action nets (Czarniawska, 2004). As Callon (1998, pp. 8–9) explained, "The agent is neither immersed in the network nor framed by it; in other words, the network does not serve as context. Both agent and network are, in a sense, two sides of the same coin ... which amounts to replacing the two traditionally separate notions of agent and network by the single one of agent-network". Thus, a firm in a region may be seen as a sociotechnical action net, comprising heterogeneous resources (such as customers and suppliers, material components, knowledge and skills, organizational routines, financial and research institutions, conventions and discourses) that provide a fabric for action (Garud *et al.*, 2010; Hernes, 2007).

Path creation does not understand agencies to be constrained by the past actions nor is path creation a rejection of the past. It is constituted by the sociotechnical arrangements that shape temporal dynamics of projects as actors frame issues about the future, coordinate their actions in the present and make sense of what may have transpired in the past. Examples include Edison's use of existing gas installations to compose the new world of electricity (Hargadon & Douglas, 2001) and efforts by Danish wind power pioneers to mobilize the electrical grid as an ally for wind power.

Path creation entertains the possibility of multiplicity of experiences associated with participation in action nets. As the journey unfolds, actors may develop new identities and frames of reference that allow them to imagine new future states and mobilize the past in new ways to accomplish their future goals and to re-organize their action nets (Garud & Gehman, forthcoming; March, 1999). New ideas, facts and artefacts emerge and are transformed through the entanglement of humans, economic resources, materials and regulations (Callon, 1991; Garud & Karnoe, 2001; Latour, 1987).

We explore path creation processes by examining the emergence of the Danish wind turbine cluster, now a dominant one in the world. It would have been difficult to imagine such a positive scenario 30 years ago. At that time, the wind turbine cluster in Denmark was described as a "cottage industry", relevant only to a small group of environmentally conscious people and comprising only a few established firms. At the same time, the capacity of one wind turbine has increased from 30 kw to 2–5 MW, i.e. from serving 8 to serve 1–3000 households with electricity. Employing observations from the wind turbine field in Denmark, the paper explicates the forming of agencies in unfolding action nets that generated the resources behind the technological cluster such as money, material components, discourses, knowledge and skills, organizational arrangements and political regulation. Accordingly, the paper focuses on resource generation processes in multiple action nets, the activation of suppliers and the framing of a market for wind power.

Brief Note on Methods

With technology studies, it is easy to report on the outcomes but more difficult to trace and map the details of the complex processes involving connections between emergent and overlapping action nets that are involved in the transformation and accumulation of resources (Latour, 1996). However, path creation suggests following practices in the making and how institutional structures are transformed even as they are performed (for a discussion on this issue in relation to regional economic change, see in particular Gertler, 2010). Indeed, it is an empirical matter to study how new visions and claims are made and how some may die out and not come into fruition. While there may be merits to models of cluster formation that "transcend the nuances of history in each community" (Powell *et al.*, 2010, p. 1), an agency and translation perspective requires the details of resource formation to give adequate insight (Callon, 1991).

One of the authors of this paper has followed this cluster for almost 30 years. There are several sources of data including (a) considerable archival materials including articles from the magazines *Wind Power Monthly* and *Naturlig Energi* (*Natural Energy*), (b) interviews with firms and actors in the emerging cluster in 1987–1990 as well as a series of new interviews in 2010 with the founders of the Wind Turbine Owners Association and (c) many books written by researchers and participants in the "social movements", including the biographies of key persons and their networks.

The Emergence of the Danish Wind Turbine Cluster

In the 1950s, Juul (a utilities engineer) developed a 200 kW wooden three-bladed upwind wind turbine with an asynchronous generator and a passive yaw that operated on the grid. Despite this accomplishment, in 1962, a special utility commission decided that the technical efficiency and cost economics associated with Juul's design were unattractive (Nielsen, 2001, pp. 11–20). However, various elements of Juul's wind turbine design informed subsequent wind turbine entrepreneurs. We describe the efforts of these entrepreneurs and other dynamics of activation and co-creation of heterogeneous resources (see Table 1 for details of key activities).

Emerging Action Nets in Design and Production

The Riisager design. In the early 1970s, Riisager (a carpenter by trade) began his experiments with renewable energy after he was called to repair an old windmill (Jensen, 2003, pp. 19–20). Using a transformed gearbox from an old military tank, Riisager came up with a 15 kW wind turbine. For blades, he used ship sails that he mounted on wooden frames. As these blades did not function effectively, Riisager invited his son, a pilot with knowledge of aerodynamics, to design a "pear-shaped" wooden blade. In an experiment with a hair dryer, Riisager discovered the now famous lift and drag mechanism wherein the aerodynamic shape of the blade creates a low pressure as it rotates thereby increasing the force of energy on the blade and causing it to accelerate.

Riisager's wind turbine caught the attention of Medelbye, an engineer whose father had worked with Juul (Wistoft *et al*, 1991: 212). Upon Medelbye's suggestion, Riisager tried an asynchronous motor (as in the Juul design) so as to connect his wind turbine to the grid. Only after operating the wind turbine for 4 days did Riisager approach the director of the local electric utility to seek his permission. The director refused for technical and legal reasons.

However, a journalist who lived near Riisager's house mistakenly reported that the utility had granted its permission. Other journalists saw this article, visited Riisager house to see his wind turbine and then wrote about it. Because of all the publicity, the utility eventually granted Riisager permission to connect his wind turbine to

Table 1. Summary of critical activities in the emergence of the Danish wind turbine cluster

	Design and production	Suppliers	Users and market	Policy and regulation	TRC
1940–1962	FL Smidth, Juul		Industry, utility	Nuclear power is the future	Helge Pedersen
1974	Riisager NIVE TVIND			Energy Plan'76	
1976		KK Electronic LM Blades Components	Riisager unlocks grid DEF Guidelines		
1977	Herborg	Økaer Blades	Møller reports on grid connec-ted turbine	Large-scale wind turbine programme	
1978	Vestas		Farmers and cooperatives buy wind turbines	Small-scale wind turbine programme	TRC begins, Helge Pedersen director
1979	Nordtank		DWO established	Subsidy scheme (30%) of investment, approval procedure	Re-sets TRC interaction with industry
1980	Bonus	Alternegy			
1981	DWT established	Suppliers continue	Slow sales in Denmark	Enforcement of policies for wind power	Load paradigm and approval criteria
1982	Search for new markets	New suppliers due to sourcing strategy and export growth	Sales to California Export boom	Extra subsidies for wind farms	Wind Atlas method
1985			DK wind farms	New payment scheme	
1986			Export slows	100 MW agreement	
1987	Number of firms increases		Utilities direct buyers		
1989			New export markets begin	Invest. subsidy removed	

the grid (Møller, 1978). This incident, in turn, led to the emergence of the first set of guidelines by the Danish Electrical Utility Association for connecting wind power to the grid.

Tvind design. Another early Danish effort emerged in 1974 in the small town of Ulfborg at the local Tvind Schools, an educational institution established to foster socialist values-based education (Jensen, 2003, pp. 60–63). The leader of the wind turbine project at Tvind contacted Copenhagen Airport to see if they could help with the design of the blade as Tvind teachers did not possess the necessary skills. The project leader was redirected to Helge Pedersen, an aeronautic engineer at Risø, the Danish nuclear power research centre (Jensen, 2003, p. 62). Helge Pedersen had designed blades for a series of F.L. Smidth wind turbines during WWII. When approached by Tvind, Helge and two of his colleagues negotiated from Risø time to work on the Tvind project. The Tvind project led to important principles for blade design and manufacturing including a technique for folding fibreglass that was then fixed with the steel structure of the blade. Tvind's efforts also led to a realization that three-bladed designs would be able to balance loads better than two-bladed designs.

Herborg design. Another noteworthy contribution was from Jørgensen, a skilled mechanic who lived in Herborg (located near Tvind). Jørgensen wanted to build a wind turbine to produce electricity for his workshop to mitigate against the high cost of energy. In contrast to Riisager, Jørgensen used his preference for metal to design a multi-bladed wind-rose design based on drawings and pictures as well as his own sense of component dimensions.

In February 1978, engineering student Henrik Stiesdal from the alternative energy movement visited Jørgensen and convinced him that a rotor with three aerodynamically shaped blades would be superior to a multi-bladed wind rose. Following a newspaper article about public funding of "new ideas for products", the two filed an application for a new Stiesdal design in spring 1978. With modest funding, they were able to design a three-bladed wind turbine with an asynchronous generator. Grove-Nielsen, who had bought the Tvind moulds, designed the blades.

The wind turbine they assembled operated without problems for several months. But, when it encountered stormy winds, the wind turbine accelerated out of control. Rather than give up, Stiesdal and Jørgensen used this incident to learn about the weak points of their design. Grove-Nielsen even published a paper titled "we can always use a storm" in Natural Energy, February 1982. By November 1978, the wind turbine had new blades and a tip-brake system. This time, the wind turbine worked successfully.

The action nets around Riisager, Tvind and Herborg were all located close to one another and near Vestas, another entrepreneurial firm (see map in Figure 1).

There were other firms and institutions in the region that experimented with wind turbines including North Jutland Institute for Renewable Energy (NIVI). Inspired by Juul's design, numerous experiments were conducted by several wind turbine start-ups (Maegaard, 2000, pp. 67–69).

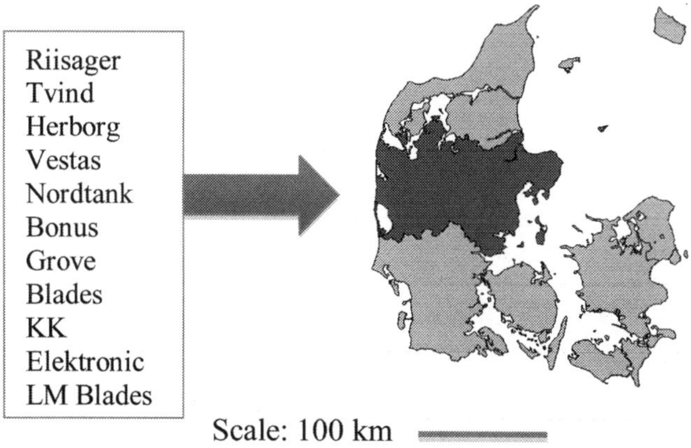

Figure 1. Location map for the region in Jutland where the wind turbine cluster emerged, except from the TRC which was established in Roskilde on Sealand.

Activation of Users and the Emergence of a Market

Møller, an award-winning journalist, was the first to buy a Riisager wind turbine in early 1976 for his countryside home in Jutland. His motivation to buy a wind turbine was to safeguard against future oil prices. He sought permission to grid-connect his wind turbine; however, there were no guidelines at that time.

In the meantime, "haunted by the press", and forced by the "growing private interest in wind energy, the Association of Danish Electric Utilities (DEF) pressed the national Price and Tariff committee to come up with rules for connecting wind turbines to the electricity grid" (van Est, 1999, p. 78). Guidelines granting independent owners the right to connect wind turbines to the local low-voltage electricity distribution grid did emerge in 1976. As per these guidelines, owners could set the price at which the utilities had to buy the electricity generated by the wind turbine; both had to share the costs associated with grid connections.

After Møllers 22 kW wind turbine was grid-connected in November 1976, a series of widely read articles began appearing in the media promoting the "new recipe for how users could make money on private investments in wind power". These articles generated a "rolling snow-ball" effect as they contributed greatly to the interest in wind power (Møller, 1978, pp. 43–46). Early owners were mostly farmers who came together to form cooperatives. These initiatives confronted resistance from local utilities and municipalities despite the DEF'76 guidelines. In addition, many users continued to be concerned about their investments because of failures. For all these reasons, the market for wind turbine remained modest at best—less than 150 wind turbines between 1976 and 1978 (Table 2).

In 1978, the Danish Windmill Owners Association (DWO) was instituted in Møller's house. Comprising about 40 people, the mandate for the DWO was to secure the investments of members with the new and unknown technology (Tranaes, interview, March 23, 2010). Six domains of interactions were targeted: (1) with politicians to create a regulatory framework; (2) with wind turbine producers to make design improvements; (3) with electrical utilities association to establish payments and conditions for grid-connection; (4) with insurance companies to cover component failures; (5) with

Table 2. Industrial activity in the Danish wind turbine cluster 1974–1989

	1974–1979	1980	1982	1985	1987	1989
Number of firms	1 to ≫6	10	10	8	11	14
Sales in Denmark (1$/ 6 dkr)	Na 200 wind turbines	3	10	30	30	70
Export sales (1$/ 6 dkr)[a]			5	350	65	65
Wind turbine kW and rotor[b]	15–30 kW (4–10 m)	45–55 kW (12–16 m)	55 kW (14–16 m)	75–100 kW (17–20 m)	100–180 kW (20–23 m)	180–450 kW (23–35 m)
Employees in cluster		50	200	3300	900	1200

Source: Various tables in Karnøe, 1991.
[a]1$ = 0.8 eurocent. [b]Dominant wind turbine size in the Danish market by rotor diameter measured in meter, a key measure for wind turbine size.

consultants to seek assistance for individuals and cooperatives in making investments in wind turbines and (6) with the government to provide national planning tools and regulations for local siting of wind power such as landscape classification (without a classified landscape, municipalities could not give permission to install wind turbines).[1]

By 1975, different concerned groups involved in wind power had initiated "wind meetings", an important collective forum for mutual knowledge sharing among producers, users and self-builders. These "wind meetings" held four to eight times a year offered a forum for the exchange of ideas between entrepreneurs, regulators and utilities. In addition, interactions between early wind turbine entrepreneurs and users created a strong environment for the coupling of learning-by-using and learning-by-doing resulting in innovations such as a double brake system. A monthly publication called Natural Energy detailed the performance of different brands of wind turbines. The location, wind speed, operational hours and component failures for each wind turbine were all reported, stimulating mutual learning in the wind turbine cluster. Moreover, the data that emerged became important later in negotiations on payments and subsidies with utilities and politicians (Tranaes, interview, March 23, 2010).

Policies and Regulation

In the 1960s, nuclear power was the preferred technology of future in most energy policy discourses. The oil crisis of 1973/1974 brought energy to the forefront (Jamison and Læssøe, 1990, p. 90, from van Est, 1999, p. 70). For instance, concerned groups emerged to problematize the then-existing discourse on energy. However, the DEF claimed that "nuclear power was the only realistic alternative to fossil fuels". The director of the DEF even commented, "you may discuss as much as you want but nuclear power you will get" (Karnøe & Buchhorn, 2008). In response, a new organization—Organization for Information about Nuclear Power (OOA) —emerged. This organization played a key role later in challenging nuclear power in the energy policy discourse over time.

New initiatives for alternative energy sources were funded by the 1974 "Energy Information Campaign" (Meyer, 2000, p. 87; Nielsen, 2001, p. 70). Shortly thereafter in 1975, the social democratic minister formally closed this initiative in order to support nuclear power. However, momentum that had already built up with these initiatives and they could not be stopped. They yielded many assets such as books and movies depicting different realities, problems and solutions that, in turn, fuelled more debates and experiments.

Nybroe's practical guide to building "Sun and Wind" (1975) is illustrative. Because of this effort, American scientist Amory Lovins presented his groundbreaking book "Soft Energy Paths" (1975) (Meyer, 2000, p. 77) that he presented in the Danish Energy Agency. Further, the Network of Energy Offices, an initiative organized via Folk high schools, spread around Denmark with the goal to formalize systematic education in building wind power plants, biogas, solar heating, etc. (Beuse, 2000, p. 62).

An important initiative emerged around experts inspired by the Club of Rome and Limits to Growth (Meadows *et al.*, 1972) when the president of the Academy of Technical Sciences (ATV), physics professor Niels Meyer, established two wind power committees (in October 1974 and September 1975). President Meyer hoped that the reports would spur wind power. Two wind power reports were published: one in May 1975 and the second in January 1976. The second ATV report recommended two programmes: a large scale (more than 30 m rotor diameter) and a small scale (less than 15 m rotor diameter) (van Est, 1999, pp. 76, 310, fn. 21).

Despite these reports, wind power was not a part of the first official energy plan. Instead, the first Danish Energy Plan EP'76 depicted a future with four to six nuclear power plants by 1995. This plan was immediately criticized by OOA. The parliamentary Energy Policy Committee asked the minister to draft a plan without nuclear power. The minister refused (van Est, 1999, p. 74). This motivated the formation of a group of scientists (including Professor Meyer from ATV) to draft an "Alternative Energy Plan" (AE'76) in summer 1976. Utilizing assumptions on economic growth similar to those in the official EP'76 plan, new techno-economic calculations emerged that served as the basis of strong media discussions, parliamentary debates and hundreds of other debates arranged by OOA and the Organization For Renewable Energy (Meyer, 2000, p. 80).

The situation changed, and large-scale wind power was officially incorporated into the Danish Energy research programme in fall 1976. Building on the ATV-report 'for large scale wind power', the \$30 mill Wind Power Program was run by the Ministry of Trade and DEFU (Danish Electrical Utilities Research unit, which took over Juul's Gedser-mill) to investigate the possibilities associated with large-scale wind power (Nielsen, 2001, pp. 108–9). However, this initiative was seen by many as a tactic to delay the introduction of wind power while nuclear power advanced (van Est, 1999; Møller, 1978, p. 38).

At this point in time in 1976, there were no plans to support small wind turbines. Pedersen from Risø was the main driver behind small-scale wind power. Inspired by efforts in the US, Pedersen advocated the creation of a test station (TRC) for small wind turbines. In 1978, his efforts resulted in a \$1 million grant to initiate a 3-year programme for small-scale wind turbines (Meyer, 2000, p. 97; Nielsen, 2001, p. 114). Only few believed that small wind turbines could contribute significantly to Danish electricity generation.

Test and Research Centre

Given the temporary nature of the grant, TRC's approach was to serve as a resource for firms with the hope that, if the industry grew, so would the TRC. Petersen, the designer of the blades for the Tvind machine, became its first director. He hired three engineers who had worked on the Tvind and Juul's wind turbines as students. Comprising theoretical thinkers and hands-on engineers, TRC researchers specialized in aerodynamics, structural dynamics and other practical engineering bases of knowledge. They had to learn about the more holistic properties of wind turbines since existing engineering knowledge was poor. However, despite the general openness within the network of wind turbine practitioners, many firms did not want the TRC looking into their specific products and design rules. In this context, TRC worked very hard to cultivate trust with practitioners in the field. For instance, to attract designs to evaluate, they offered an award for the best wind turbine design installed at Risø's site.

In the late 1970s, political negotiations emerged around subsidies for wind power. Several actors including the Ministry of Trade, the media as well as industry associations contested this subsidy scheme. However, the scheme was passed in June 1979 with the Minister of Housing arguing that "the 30% subsidy was not meant to support basic development work", but instead for "creating production opportunities for the Danish industry in such a way that stable batch-production could be achieved" (van Est, 1999, p. 79).

In its passage, the new government regulation required firms to seek TRC's approval of their wind turbine designs so that buyers could get subsidies. This requirement forced the industry to open its doors to the TRC. It is important to note that the institution of the TRC did not stem from a Danish industrial or technology policy tradition; indeed, it was a major institutional innovation that not only stimulated the market but also induced actors in industry and those in institutional bodies such as the TRC to collaborate. Over time, TRC researchers were gradually accepted by the industry as TRC's positive influence in the creation of the cluster became evident.

As interactions with industry progressed, TRC researchers found that they did not possess all the knowledge required to create robust criteria for the evaluation of wind turbines. Consequently, TRC researchers began interacting heavily with early wind turbine users and manufacturers. Criteria were borrowed from the DWO whose members demanded safety; these were supplemented with other structural criteria used in buildings. Because of the uncertainties involved, these criteria were on the conservative side, encouraging heavyweight constructions in the tradition of machine shops. Although not anticipated, such heavy weight constructions reduced the structural vulnerability of the Danish wind turbines when they encountered unexpected high dynamic wind loads (Stoddard, 1986).

TRC's rules were flexible and continued to evolve as wind turbine manufacturers suggestions were incorporated into the overall framework (Lundsager & Hjuler Jensen, 1982). TRC did not think of itself as conducting "ivory-tower" research; instead, it tried to bridge different knowledge communities. Sharing knowledge about wind turbines through early reports put out by the TRC occurred as much through "qualitative stories" and "metaphors" as through the precise language of mathematical formulations.[2] These reports as well as the interactions between TRC researchers and manufacturers resulted in stabilizing wind turbine designs and greatly enhanced their performance. For instance, the performance of the 55 kW design improved more than 50% from 1981 to 1984 (Karnøe, 1991).

Renewal and Continuity of Design and Production

The 1979 subsidy-approval scheme re-set the economics of wind power. Subsequently, Vestas, Nordtank, Bonus and Micon all decided to diversify into wind turbines by licensing and using elements from Riisager and Herborgs designs. Over time, these firms became key players in the global wind turbine field. By 1980, about 10 firms were actively selling wind turbines in Denmark.

Around 1979, knowledge about who was doing what with wind turbines was "in the air" in the region. For instance, word that Herborg wind power could not fund the production of 50 wind turbines became common knowledge. In the meantime, Stiesdal (of Herborg) heard that Vestas had not been successful with its vertical Darrieus-turbine design. Discussions ensued between the two companies, resulting in a license agreement that transferred Herborg's design to Vestas.

Similarly, Nordtank suffered declines in sales of their traditional products: oil tanks and watering equipment for farmers. A production engineer convinced Nordtank's owner to go into wind power (Jensen, 2003, p. 78). The project team used design solutions from designs pioneered by Riisager and Herborg including blade designs from Grove-Nielsen and Økær Energy who were Herborg's blade suppliers. Nordtank's newly re-invented and re-designed wind turbine was presented at Farmers Fair in May 1980.

Bonus too was suffering declines in the sales of its traditional products. Bonus' owner did not want to spend money on "loose ideas" proposed by his son, a skilled mechanic. But after seeing Nordtank's wind turbine at the Farmers Fair, the son was able to convince his father. Subsequently, Bonus re-engineered a Nordtank wind turbine (Jensen, 2003, pp. 82–83) and obtained design knowledge from NIVI and Risø test centre. Bonus made the metal frame for the nacelle of the wind turbine (the housing for the generating components) and activated local suppliers to supply components such as bearings, blades and main shaft.

All these companies built upon and transformed their existing skills and routines for product designs and manufacturing. They had a hands-on approach to design and development based upon a mix of skilled mechanics, "teknikum-engineers" and a few engineers from Technical Universities. Knowledge for the design and construction of wind turbines gradually became embedded in new shop floor routines and practices as competencies and design rules began accumulating through practical experimentation and learning-by-using from operational experience. To the people involved, the wind turbine presented itself as a set of problems similar to those they had confronted in the construction of machines that demand the integration of different parts, and the Danish wind turbines became known as "heavy-weight". During the 1980s, the work method went from an informal process of design and experimentation with a low degree of knowledge explication to a more formalized process involving formalized R&D interacting with manufacturing and use.

Supplier Activation

Supplier competencies did not pre-exist and had to be built over time in collaborative network settings that encouraged and created mutual knowledge sharing. Pioneers such as Riisager and Herborg Wind Power activated key suppliers for electronics, blades and bearings from the local industrial district. Local suppliers were specialized skill-containers who transformed their skills in order to make "just another component for a customer" (Andersen & Christensen, 1999). For example, Riisager activated and enrolled KK Electronics to his action net. KK Electronics had 20 employees in 1976 and supplied various electronic control board systems. Upon receiving Riisager's order, KK developed a simple relay control and improved the hardware and critical measurement points and control loops. KK Electronics eventually became the market leader.

For blades, Riisager approached LM Fiber Glass, a supplier to boat manufacturers, to see if it might design blades for wind turbines. But, LM was sceptical about wind power and did not pursue this opportunity. With the California Gold Rush, however, another company, Coronate Boats (re-named to Alternegy), became involved. Alternegy took over the finished product that Grove-Nielsen and his company Økær Blades had developed between 1975 and 1980 in interaction with Herborg and others.

Components such as gears and generators were not available in Denmark and had to be sourced from suppliers in Sweden, Finland and Germany. During the 1980s, standard components became more specialized. All the major wind turbine manufacturers collaborated with their own suppliers to design their own special components. For each new and bigger wind turbine design, wind turbine firms had their own "gear-engineer" located for months at the supplier's premises. Overall, the wind turbine cluster of suppliers emerged as local firms responded entrepreneurially to their declining business situation by exploring new opportunities with wind turbines.

There were overlaps in the suppliers for Vestas, Micon, Nordtank and Bonus. However, to avoid dependency on any one supplier and to ensure quality and prices of the parts, each firm had a second source. In contrast to Micon and Bonus, Vestas began internal design and manufacturing of blades and electronics so as to control critical component knowledge in-house for certification purposes.

Supplier relations followed the institutionalized practices of the industrial region – a mix of competition, collaboration and sharing risk (Andersen and Christensen, 1999; Lorenzen, 2003; Maskell *et al.*, 1998). In addition, the industrial region was typical for the "Danish" industry structure composed of collaborative SMEs (Karnøe *et al.*, 1999). Such a structure offered wind turbine manufacturers an opportunity to exploit external economies (Langlois, 1992) rather than create specific technological competencies in-house.

International Users

From 1980 to 1982, the home market grew slowly with total revenues of $10 million for the 10 manufacturers. However, the learning experiences from 1974 to 1981 prepared the emerging cluster well for a new growth opportunity that emerged in California. The Danish Wind Turbine manufacturers association funded the creation of a report on the California market in spring 1982. This report highlighted a new market in California—the "1980 programme" aimed at obtaining 10% of the energy from wind power before 2000. There were federal and state tax credits for investors and electric utilities would pay a 10-year fixed price for wind power (via the PURPA).[3] Noticing this opportunity, all big Danish manufacturers went to California to establish export opportunities.

The number of Danish wind turbines exported annually increased from 30 to 2000 per annum between 1982 and 1985. Industry employment grew from 300 to 2500, making the "ugly duckling" (i.e. the wind turbine) famous in Denmark. The Danish market share in California increased from 0% in 1981 to 65% in 1985, while the market expanded from 71 to 1250 MW in California. Although the large weight of Danish winds turbines created a legitimacy problem in the home market (see below), the designs outperformed their lightweight American counterparts (Stoddard, 1986). Also the strong dollar favoured Danish export and profits.

This California "gold rush" generated tremendous organizational challenges even as it offered growth opportunities for the SMEs (Karnøe, 1991). It jump-started (a) the creation of new routines as "large batch production" was undertaken, (b) sales of wind farms to developers, (c) product guarantees, (d) sourcing and (e) shipping logistics. "Big money and stakes" led insurance firms and certification companies to demand explicit criteria and the explication of knowledge used in the design of turbines before granting any funds and before extending insurance coverage. The load paradigm and Risø researchers helped to legitimize the highly tacit knowledge. Also, in servicing the export market, wind turbines took on a new meaning when wind turbine manufacturers understood the magnitude of the efforts required to sell and service a capital good with a lifespan of 15–20 years with 3–5-year product guarantees. New procedures for service, repair and guarantees had to be instituted through a costly process of trial-and-error learning.

Regulations and Domestic Users in the 1980s

The market in Denmark grew slowly in 1980 despite the subsidy scheme. In response, left-wing political parties used the complaints from DWO to convince the Minister of Trade to enforce the DEF'76 guidelines. Another resource for the wind turbine cluster came from the meteorologists at Risø nuclear centre. In 1981, Risø meteorology researchers invented the reliable Wind Atlas method for calculating the wind speed given the landscape-type ruggedness (Nielsen, 2001). This became an important calculative device to reduce the uncertainties associated with the nonlinearity involved in the transformation of wind into energy (Møller, 1978).

The booming export market in California had a positive effect on the domestic market as it shaped expectations and market policies at a point in time when the existence of wind power in Denmark was fragile and contested. The national Energy Plan of 1981 mentioned that wind power could contribute 10% electricity by 2000, but it was still a nuclear power energy plan. However, the '10% number' was enacted more optimistically by chairman Madsen of WTM (Nielsen, 2001). Optimism was important, as critics and skeptics contested wind power strongly so as to negatively influence adaptive expectations about this source of energy.

In one debate centred on the size of the 22–30–55 kW wind turbines, opponents claimed that it was "too small to matter energy production wise" in comparison to the large-scale wind power programme. Intense criticisms came in a series of articles from a member of the Nuclear Power Promoting Association in Denmark and a contributor to the overall Danish energy policy (Elbek, 1980a–d). Madsen, the chairman of WTM, responded and offered counter-evidence from recent performance data (Madsen, 1981). Despite this evidence, though, many had doubts about the potential of this seemingly low-tech system and the small-scale programme.

Reports on the California Gold Rush offered the red-green political coalition arguments to save wind power. Wind power policies could now be justified with reference to export and new jobs. During the 1980s, the activities to stabilize wind power in the energy policy discourse were intensified by concerned groups. OOA's activities continued after the Parliament voted nuclear power out of the Danish energy policy in 1985.

The California wind power programme also legitimized wind farms resulting in new direct investment subsidy (25% plus an additional 15% for large-scale, privately owned wind farm installations). In addition, the economics of wind power improved by a regulation in 1984 that phased out the investment subsidy (subsidy gradually reduced and ended 1989) and replaced it with a fixed subsidy per kWh produced (a forerunner of the feed-in tariff model). The economics of wind power was secured with the 1984–1994 agreement with electrical utilities, and users had now a predictable investment horizon. However, the growth of wind farm installations and profits de-legitimated the public subsidy programme because it allowed rich investors to operate wind farms without involving local inhabitants. Public funding of wind farms that would yield large profits to the rich was not considered to be a legitimate use of public funds (K. Worsaae, Director in Energy Agency, personal conversation, 1996). Further, such speculative investors were not wanted by DWO as they wanted to maintain political and public legitimacy in the subsidies (Tranaes, interview, March 23, 2010).

Consequently, a new governmental act from December 1985 imposed restrictions on private ownership of wind turbines. Bypassing the norm of involvement of WPO and WTM, this act placed geographical boundaries on ownership (the geographical distance between an owner's residence and the wind turbine and the two locations had to be in the same municipality). The second restriction limited the extent to which individuals could invest in wind power relative to their use of electric power. Specifically, the act stipulated that the share of wind power must not exceed one's electricity consumption by more than 35%.

This act led to losses of sales in the private market and resulted in a crisis in the industry as many contracts were canceled. But all was not lost. As part of this act, energy authorities sought an agreement from the electric utilities that they would install 100 MW of wind power energy from 1986 to 1990. This represented a substantial increase in the home market as private installations from 1978 to 1985 totalled only 75 MW. Moreover, this agreement had the effect of establishing the electric utilities as a new group of direct users and a resource for wind power development, as they had to invest in wind power.

The "100 MW act" is a striking example of the emerging regulatory capacity to steer, intervene and nurture the emerging industry. From a regulatory angle, the governance regime from the 1970s was upgraded, and the Energy Agency monitored and 'steered" the cluster. Besides granting wind power the right to exist, and give wind power electricity priority in the grid, it also involved continuous monitoring the fairness and legitimacy of subsidies for wind turbine owners, and the coercive 100 MW act (Hansen, 1985) forced the electric utilities to make direct investors in wind power from 1986–90.

In this way, the Danish governance regime for organizing and monitoring the regulations was an important part of the cluster success. It facilitated experimentation and learning and was done in close interaction with those involved in the emerging techno-economic network of the wind turbine cluster. Another example is the Renewable Energy Committee (within the Danish Board of Technology) sponsored small-scale entrepreneurial activities from 1982 (a total of $45 million in smaller grants were allocated during 1982–1991). It became a forum for generating legitimacy for renewable energy and wind energy, as the chairman Professor Niels Meyer, former chairman of ATV, had a network of contacts with ministries and politicians (Meyer, 2000). However, the more controversial regulations were linked to the investment subsidy scheme, the new 1984-subsidy scheme for each kWh wind power generated electricity, as they violated the neoliberal industrial policy tradition of not intervening in the market process.

The regulations were dependent upon a stable red-green political coalition in the Parliament, which annually had to fight for a majority to renew the subsidies on the state budget.

Further, despite national landscape planning and allocation of sites for wind turbine installation, local municipalities often stalled the process, and existing regulations were dependent upon political enforcement to be implemented. Such delicate regulations were not based upon a strict bureaucratic approach, but were flexible and involved compromises that were reached in committees and forums where involved actors had legitimate seats. This governance regime was full of contestation, conflicts, pressure and compromises but it managed to preserve the commitment to continue with wind power in the Danish energy system in these formative years (Karnøe & Buchhorn, 2008).

A New Beginning

By 1986, the California tax credit market subsidy had expired. In conjunction, there arose an unexpected need for repairs of the Danish wind turbines that had been installed in California. Moreover, there was reduced trust in the California market of wind turbine technology in general and of Danish wind turbines in particular (Gipe, 1995).

The decline in the California market was far from offset by growth in the Danish and other international markets. Attracted by the growth opportunities, the number of wind turbine producers increased to 14, but six firms had serious economic problems and several went bankrupt. However, firms often "reappeared" and the remaining firms often hired employees laid off from other Danish wind turbine firms. In this way, competencies and knowledge were retained in the field.

During this slump, wind turbine companies continued to believe in their technology and made new generations of 180–450 kW wind turbine designs. The Danish home market recovered in 1988, and the recovery was, for the first time, driven by increases in both private installations and utility installations. To implement the agreement to install 100 MW stipulated by the 1985 act, the utilities had to develop organizational competencies and identify relevant sites. By the time these issues could be sorted out, it was already 1988. This delay was "lucky" as a home market sales boost (at a time when it was most needed) allowed the utilities to access new and improved 180–450 kW wind turbine designs. With these efforts, the Danish wind turbine industry was ready to dominate the 1990s in new markets like Germany, UK and Spain. Indeed, the Danish wind turbine cluster took advantage of the first mover position from 1972 to 1990, and became globally leading. From late 1990s newly established competitors like Spanish Gamesa (started as Vestas joint venture in 1992) and Indian Suzlon located their R&D facilities in Denmark to buy into and recruit high-level wind power-specific engineering and business expertise. Siemens bought Bonus Wind Power in 2005, and by 2006, a report from the Ministry of Science saw Biotech, IT and wind power as new Danish high-tech industries.

However, this world leading position was only possible due to regulatory continuity that almost was lost when the supportive red-green majority lost power in the 1991 election. The new ministers for environment and energy did not support wind power installations, although their rhetoric claimed otherwise. The overflows from using wind turbines, such as noise and visual landscape pollution, were mobilized against wind turbines, and new guidelines for classification of the landscape fostered great uncertainty among producers and users of wind turbines in Denmark. Private Danish installations of wind turbines almost stopped in the years 1992–1993; only utility installations remained. An unexpected political scandal got the red-green coalition back into government with a majority in 1993, and the new Minister of Energy and Environmental Affairs, Svend Auken, declared his enthusiastic support to combat climate change with renewable energy, witnessed by a new Energy Plan (E-21, focusing on CO_2 reductions). A new feed-in payment regime was negotiated with the state, the wind turbine owners and the electrical utilities, and this led to increased private installations. From 1994 to 2000, and jointly with some utility-owned off-shore wind farms, wind power reached a record high 18% share of Danish electricity generation (Karnøe & Buchhorn, 2008).

Discussion

This paper attempted to show how the notion of agency in path creation in the emergence of clusters. To do so, the paper covers two periods in the development of wind turbines in Denmark: 1972–1979 and 1980–1990. Important activities and regulations that facilitated the transformation of the existing cluster of firms and the emergence of the wind turbine cluster are overlooked if the focus is only on the period between 1980 and 1990 as is the case with many studies. In the first period new agencies were formed as multiple action nets enacted the "1974-energy crisis" and a dominant coalition promoted nuclear energy as the only solution. The emergent action nets that mobilized wind power as an alternative were highly different and counted both concerned social groups, entrepreneurs like Riisager, the Tvind action net, academics, and politicians. Wind power became an occasion for reorganizing agencies and competencies as well as an occasion for reorganizing policy regulation. The emergence of the wind turbine cluster cannot be seen without this transformation served as an alternative with the reactivation of Juul's design, transformation of suppliers and the emergence of users and buyers of wind turbines. Building on this, new company entrepreneurs built upon Herborg and Riisager's pioneering work and transformed existing competencies and routines to exploit the California Gold Rush during the second period, 1980–1987. Further, an emergent new political coalition behind the regulatory market framework was able to layer, convert and recombine (Martin, 2010) regulatory instruments to "steer" (Kemp *et al.*, 2001) the development of the wind turbine cluster (such as inducing utilities to implement rules that enabled wind power to be added to the centralized electricity system). Also, the "100 MW agreement" converted electrical utilities into direct buyers and allowed for the sales of new and upgraded designs in the Danish home market 1988–1990. When the new European market was created, Danish manufacturers were ready and became world leading in the 1990s.

The Danish wind turbine cluster originated with sense-making that occurred within multiple localized action nets that led to new techno-economic projects. Different participants understood these issues differently depending upon their frames and action nets. Unlike the path dependence perspective of lock-in, actors were not constrained by their action nets. We saw how their specific experiences induced them to develop new identities and frames of reference that allowed them to imagine new future states and to re-organize their action nets by translating and enrolling others into these new projects. Such shifts in agencies associated with technical innovations can only be studied by attention to details, and this contrasts the type of explanations of institutional configurations by national innovations systems.

Indeed, the notion of path creation makes it possible for us to understand that agency is not constrained by structures that have emerged from the past. Path creation understands embedding structures as offering dispositions to act and a capacity to articulate and frame issues. Ascribing only inertia to existing arrangements provides an impoverished notion of agency, as both maintenance and transformation are outcomes of work and struggle. Different actors in the Danish cluster built upon and transformed their own existing competencies and built upon and transformed competencies from suppliers in the region so as to create a wind turbine.

As ready-made resources and a market for the wind turbine technology were not pre-existent, pioneers and company entrepreneurs confronted several challenges in their efforts at mobilizing internal support for funding and for developing competencies. As with producers and suppliers, markets, regulations and users too did not pre-exist. For users to become a valuable resource for an emergent cluster, we have to appreciate how a market framework for wind power emerged. A political "plan" for R&D support and market subsidies for small-scale wind turbines was not part of the first official energy plan. Translations were involved to set processes in motion by transforming dis-interested actors and mis-fit materials. These translations stemmed from local acts, and the eventual structuring and relative stability of the market framework depended upon the coming together of heterogeneous elements such as users, electrical utilities, political parties, insurance companies, classification schemes, legal guidelines and calculative devices. The process of translating wind power to become something and nuclear power to become nothing was full of resistance, conflicts and struggle for legitimacy. Central in this process is that the action nets that provided existence to wind power gained more and more strength while the action nets that formerly provided existence to nuclear power weakened. Moreover, the process entailed an interweaving of discursive, procedural, and technical and calculative elements.

The industrial cluster emergence occurred through co-creation and interweaving of resources across multiple arenas of use, production, regulation and evaluation (for example, when Medelbye saw Riisager's wind turbine and suggested an 'asynchronous generator used by Juul, or when Stiesdahl heard about Vestas experimental failure and offered them Herborgs wind turbine).

The co-creation also cut across the involved sociotechnical elements in the action-nets, as the emerging action nets involved several nonhuman materials as well. For example, Medelbye was attracted to Riisager by seeing the material wind turbine as he drove by; blade failures in Herborgs action-net were referred to as stimulating re-thinking and re-design, and the storm was seen as a very helpful actor by blade designer Grove-Nielsen. Further, Vestas failed experiments was part of setting their interests to be ripe for Stiesdals translation when he attempted to make Vestas take over Herborgs order of fifty wind turbines, as Herborg could not get finance in the bank. This process of co-creation and interweaving across arenas and sociotechnical elements stemmed from a process that appears disorderly.

More importantly as the process unfolded, the learning dynamics were subject to the critical details of timing and sequencing of events that are associated with increasing returns thinking. For instance, (a) storms showing weaknesses in components, (b) unemployment becoming a justification for market subsidies and (c) the Californian market for wind power allowed for volume-based learning, profits and competence upgrading. In processes as complex and nonlinear as cluster emergence, "garbage can" as a metaphor from organizational theory (Cohen *et al.*, 1972) may provide an apt conceptualization for understanding how different solutions and problems become connected.[4] The central idea here is that a link between actors, problems and solutions depends heavily on the simultaneity of their "arrivals", and that the interweaving and ordering of resources may stem from disorderly processes. If learning in clusters occurs as a path is emerging such that learning dynamically changes the trajectory of the path, then we are far from predictable learning dynamics (Martin & Sunley, 2003).

This account of the emergence of the Danish wind turbine cluster shows that it did not result from a public policy strategy that picked the winner. Nor was it shaped by pure pre-existing market forces. Clusters based on new technologies do not offer clear-cut valuations *ex ante* to be qualified as economically viable. Yet, Danish regulators stepped in with subsidies and grants to nurture and shape an emerging technology. At the same time, Danish regulations have witnessed near-failures with biogas and realized only modest success with solar panel technologies. In other words, there is no Danish recipe.

However, the implication is not to stop influencing cluster formation. Cluster-forming initiatives such as defence contracts in Silicon Valley, Korean industrial development, Chinese development of wind power and German solar power suggest the importance of fostering distributed and emergent actions that allow for the interweaving of resources and the transformation of existing resources over time. Public policy may have a new role in fostering a fertile ground for such path creation-type processes to unfold. It can be through the creation of "hybrid forums" (Callon *et al.*, 2009; Garud & Gehman, forthcoming) that foster interactions and dialogues between actors, artefacts and other resources so that clusters can emerge. In the Danish case, a central governmental body did not control and steer this cluster in a bureaucratic manner. Instead, we see the Energy Agency (and the Renewable Energy Committee) engaged in a flexible process of 'steering through monitoring" and participation in hybrid forums. This may also facilitate gradual change through layering, conversion and re-combination of regulatory instruments (Martin, 2010) as was the case with Danish wind turbines. In these forums, concerns and issues related to the emergent cluster were constantly raised and addressed in ways that preserved enough trust, legitimacy and commitment for all involved parties to co-create cluster resources in a transformative fashion.

Notes

1. Several other devices framed the economics of wind power. For instance, wind turbines were classified as a building. Consequently, wind entrepreneurs could get a 20-year loan rather than a loan for 4–6 years if it had been classified as a machine.
2. Stoddard (1986) noted that, of the 25 most important research reports of the 1980s on wind turbine designs, 12 came from the Danish TRC.
3. The Public Utility Regulatory Policies Act (PURPA), part of the National Energy Act of 1978, attempted to eliminate barriers for wider adoption of renewable energy sources in the US. Utilities were compelled to buy electricity from independent power producers at fair prices. Jointly with federal and California tax credits amounting to some 40–45%, these incentives encouraged the use of alternative energy (Gipe, 1995). In fact, tax credits were so high that profits were made even from wind turbines that generated hardly any electricity.
4. Interview with James March by Diane Coutu (*Harvard Business Review*, October 2006, p. 88).

References

Andersen, P. H. & Christensen, P. R. (1999) Internationalization in loosely coupled business systems, in: P. Karnøe, P. H. Kristensen & P. H. Andersen (Eds) *Mobilizing Resources and Generating Competencies: The Remarkable Success of Small and Medium Sized Enterprises in the Danish Business System*, pp. 205–231 (Copenhagen: Copenhagen Business School Press).

Beuse, E. (2000) Grundtvig køber vindmølleanlæg, in: E. Beuse *et al.* (Eds) *Vedvarende Energi i Danmark 1975–2000* [Renewable Energy in Denmark 1975–2000], pp. 83–88 (Aarhus :Organization for Renewable Energy). Available at: http://www.ove.org

Brown, J. S. & Duguid, P. (2000) *The Social Life of Information* (Boston, MA: Harvard Business School Press).

Callon, M. (1991) Techno-economic networks and irreversibility, in: J. Law (Ed.) *A Sociology of Monsters: Essays on Power, Technology and Domination*, pp. 132–161 (London: Routledge).

Callon, M. (1998) Introduction—the embeddedness of economic markets in economics, in: M. Callon (Ed.) *The Laws of the Markets*, pp. 1–57 (Oxford: Blackwell Publishers).

Callon, M., Lascumes, P. & Barthe, Y. (2009) *Acting in an Uncertain World—An Essay on Technological Democracy* (Cambridge MA: MIT Press).

Cohen, M., March, J. & Olsen, J. (1972) A garbage can model of organization choice, *Administrative Science Quarterly*, 17(1–25), p. 2.

Czarniawska, B. (2004) On time, space, and action nets, *Organization*, 11, pp. 773–791.

David, P. A. (1993) Path dependency and predictability in dynamic systems with local network externalities: A paradigm for historical economics, in: D. Foray & C. Freeman (Eds) *Technology and the Wealth of Nations*, pp. 209–231 (London: Pinter).

Elbek, B. (1980a) Vedvarende energi: 28 milliarder med en negativ forrentning på 4%, Ingeniøren nr. 35.

Elbek, B. (1980b) Hvor meget producerer en vindmølle? Ingeniøren nr. 12.

Elbek, B. (1980c) Torgny Møller maler for godt, Ingeniøren nr. 41.

Elbek, B. (1980d) Vindkraft og saglighed, Ingeniøren nr. 43.

van Est, R. (1999) *Winds of Change: A Comparative Study of the Politics of Wind Energy Innovation in California and Denmark* (Amsterdam: International Books).

Garud, R. & Gehman, J. (Forthcoming) Sustaining sustainability journeys: Implications for strategizing and organizing, *Research Policy*.

Garud, R. & Karnoe, P. (2001) Path creation as a process of mindful deviation, in: R. Garud & P. Karnøe (Eds) *Path Dependence and Creation*, pp. 1–39 (Mahwah, NJ: Lawrence Erlbaum Associates).

Garud, R. & Karnoe, P. (2003) Bricolage versus breakthrough: Distributed and embedded agency in technological entrepreneurship, *Research Policy*, 32, pp. 277–300.

Garud, R., Kumaraswamy, A. & Karnøe, P. (2010) Path dependence or path creation, *Journal of Management Studies*, 47(4), pp. 760–774.

Gertler, M. S. (2010) Rules of the game: The place of institutions in regional economic change, *Regional Studies*, 44, pp. 1–15.

Gipe, P. (1995) *Wind Energy omes of Age* (New York: Wiley).

Hansen, J. L. (1985) Samarbejde med Folketinget på sidelinjen, *Naturlig Energi*, 9, p. 7.

Hargadon, A. B. & Douglas, Y. (2001) When innovations meet institutions: Edison and the design of the electric light, *Administrative Science Quarterly*, 46(3), pp. 476–501.

Hernes, T. (2007) *Understanding Organization as Process—Theory for a Tangled World* (London: Routledge).

Iammarino, S. & McCann, P. (2006) The structure and evolution of industrial clusters: Transactions, technology and knowledge spillovers, *Research Policy*, 35, pp. 1018–1036.

Isaksen, A. & Hauge, E. (2002) *Regional Clusters in Europe*. Observatory of European SMEs Report 2002, No. 3 (Luxembourg: European Communities).

Jacobsson, S. & Bergek, A. (2003) Transforming the Energy Sector: The Evolution of Technological Systems in Renewable Energy Technology, in: J. Klaus, B. Manfred & W. Anna (Eds) *Governance for Industrial Transformation. Proceedings of the 2003 Berlin Conference on the Human Dimensions of Global Environmental Change*, pp. 208–236 (Berlin: Environmental Policy Research Centre).

Jensen, I. K. (2003) *Mænd i Modvind - et dansk industrieventyr* (København: Børsens Forlag).

Karnøe, P. (1991) *Danish Wind Turbine Industry—A Surprising International Success: On Innovations, Industrial Development and Technology Policy* (Copenhagen: Samfundslitteratur) (in Danish).

Karnøe, P. (1996) The social process of competence building, *International Journal of Technology Management*, 11(7/8), pp. 770–790.

Karnøe, P. (1999a) *The business systems framework and the Danish SME's*, in: P. Karnøe, P. H. Kristensen & P. H. Andersen (Eds) *Mobilizing Resources and Generating Competencies: The Remarkable Success of Small and Medium Sized Enterprises in the Danish Business System*, pp. 7–66 (Copenhagen: Copenhagen Business School Press).

Karnøe, P. (1999b) When low-tech becomes high-tech: The social construction of technological learning processes in the Danish and American wind turbine industry, in: P. Karnøe, P. H. Kristensen & P. H. Andersen

(Eds) *Mobilizing Resources and Generating Competencies: The Remarkable Success of Small and Medium Sized Enterprises in the Danish Business System*, pp. 139–182 (Copenhagen: Copenhagen Business School Press).

Karnøe, P. & Buchhorn, A. (2008) Path creation dynamics and winds of change in Denmark, in: Lafferty *et al.* (Eds) *Promoting Sustainable Electricity in Europe*, pp. 73–101 (Cheltenham and Massachusetts: Edward Elgar Publishing).

Kemp, R., Rip, A. & Schot, J. (2001) Constructing transition paths through the management of niches, in: R. Garud & P. Karnøe (Eds) *Path Dependence and Creation*, pp. 269–302 (Mahwah, NJ: Lawrence Erlbaum Associates).

Langlois, R. (1992) External economies and economic progress: The case of the microcomputer industry. *Business History Review*, 66, pp. 1–50.

Latour, B. (1987) *Science in Action: How to follow scientist and engineers through Society* (Milton Keynes: Open University Press).

Latour, B. (1996) *Aramis and the Love of Technology*, (Cambridge, MA: Harvard University Press).

Lorenzen, M. (2003) Low-Tech localized learning: The regional innovation system of Salling, Denmark, in: B. T. Asheim, L. Coenen & M. Svensson-Henning (Eds) *Nordic SMEs and Regional Innovation Systems—Final Report* (Oslo: Nordic Industrial Fund). Available at http://www.nordicinnovation.net

Lundsager, P. & Hjuler Jensen, P. (1982) Licensing of windmills in Denmark. Proceedings from the *Fourth International Symposium on Wind Energy Systems*. Stockholm, September.

Lundvall, B.-A. (Ed.) (1992) *National Systems of Innovation: Towards a Theory of Innovation and Interactive Learning* (London: Pinter).

Madsen, B. T. (1981). Vindmoeller producerer mere end forventet. Ingeniøren, nr. 23.

Maegaard, P. (2000) Nordvestjysk Folkecenter for Vedvarende Energi, in: E. Beuse *et al.* (Eds) *Vedvarende Energi i Danmark 1975–2000* [Renewable Energy in Denmark 1975–2000], pp. 67–74, 111–132 (Aarhus: Organization for Renewable Energy www.ove.org). Available at http://www.ove.org.

Martin, R. (2010) Roepke lecture in economic geography – Rethinking regional path depedence: Beyond lock-in to evolution, *Economic Geography*, 86(1), pp. 1–27.

March, J. G. 1999. Research on organizations: Hopes for the past and lessons from the future, *Nordiske Organisasjonsstudier*, 1, pp. 69–83.

Martin, R. & Sunley, P. (2003) Deconstructing clusters: chaotic concept or policy panacea? *Journal of Economic Geography*, 3, pp. 5–35.

Maskell, P., Eskelinen, H., Hannibalsson, L., Malmberg, A. & Vatne, E. (1998) *Competitiveness, Localised Learning and Regional Development* (London: Routeledge).

Meadows, D. H., Meadows, D. I., Randers, J., Behrens W. (1972) *The Limits to Growth: A Report to The Club of Rome* (New York: Universe Books).

Meyer, N. I. (2000) Politik og Vedvarende Energi, in: E. Beuse *et al.* (Eds) *Vedvarende Energi i Danmark 1975–2000 (Renewable Energy in Denmark1975–2000)*, pp. 75–110 (Aarhus: Organization for Renewable Energy), Available at http://www.ove.org

Møller, T. (1978) *Vinden Vender Andelsforlaget Vistoft* (Knebel: Cooperative publisher Vistoft).

Nielsen, K. H. (2001) Tilting at windmills, 2nd ed. Phd dissertation. History of Science Department, Aarhus University, and Department of Organization, Copenhagen Business School.

Porter, M. E. (2000) Locations, clusters and company strategy, in: G. L. Clark, M. P. Feldman & M. S. Gertler (Eds) *The Oxford Handbook of Economic Geography*, pp. 253–274 (Oxford: Oxford University Press).

Powell, W., Packalen, K. & Whittington, K. (2010) Organizational and institutional genesis: The emergence of high-tech clusters in the life sciences, in: J. Padgett & W. Powell (Eds) *The Emergence of Organization and Markets*.

Stoddard, F. S. (1986) The California experience. *Proceedings of the Danish Wind Energy Association Wind Energy Conference*.

Wistoft, B., Petersen, F. & Hansen, H. (1991) Dansk elforsynings historie 1891–1991, bd. 1. Danske Elvìrkers Forening.

Path Dependence and New Technological Path Creation in the Danish Wind Power Industry

JAMES SIMMIE

Department of Planning, Oxford Brookes University, Headington Campus, Gipsy Lane, Oxford OX3 0BP, UK

ABSTRACT *In this paper, it is argued that in order to overcome the limitations of canonical path dependence and new path creation theories that arise from the different ontological approaches of economics and sociology, a hybrid socio-economic theory is required that examines the contributions of intelligent agents to the creation of new technological pathways in conditions of path dependence. The main aim of such a theory is to explain the processes by which agents may collectively contribute to the emergence of new technological pathways and overcome the barriers confronting them as a result of the evolution of historical forces that establish the path-dependent trajectories of contemporary technologies. The arguments are illustrated empirically by analysing the roles of inventors and innovation pioneers located in economic niches together with the diffusion of new technologies to the attainment of critical mass and the creation of new pathways. The roles and processes engaged in by such actors, and the path-dependent barriers confronting them, are illustrated using the historical creation the renewable energy technological pathway in wind power in Denmark.*

Introduction

Over the last two decades or so, a growing dissatisfaction with some of the underlying assumptions of neo-classical economics, such as those of the hypothetical existence of rational average actors whose actions drive aggregate economies towards equilibrium, has led to an evolutionary turn in economic theory. Inspired by the work of the Austrian economist Joseph Schumpeter (Schumpeter, 1939, 1942), a number of distinguished economists such as Nelson and Winter (1982), Dosi *et al.* (1988), Hodgson (1993), Arthur *et al.* (1997), Foster (1997), Metcalfe (1998), Potts (2000), Fagerberg (2003), Dopfer (2004), Metcalfe and Foster (2004) and Witt (2003) have sought to understand how real economies evolve through time. Neo-Schumpeterian theories concerning the role of technological innovation in driving economic evolution have been prominent in explanations of such evolution (Witt, 1993; Dosi *et al.*, 1998; Metcalfe, 1998). Between them they have studied technological innovation at the company level, competition, diffusion and structural change at the levels of markets and sectors and growth, long waves and international trade at the macro-level. Studies such as these have established the value of adopting an evolutionary perspective to understand the historical growth and change of real as opposed to hypothetical economies.

As this is still an emerging approach in economic theory, there is, as yet, no settled view on which elements of evolutionary theory should be the standard analytical tools of analysis. So far, at least four main approaches to analysing economic change and innovation have been developed. These are the Darwinian inspired biological analogy and the notion of the co-evolution of institutions with economic change, complex adaptive systems theory, panarchy and path dependence theory.

This paper focuses on the last of these four alternatives. In particular, it seeks to address the question "In conditions of path dependence how are new technological and industrial pathways created in the first instance?" It is argued that, in addition to the standard analyses of path dependence and lock-in, we need better explanations of how new technological and industrial pathways are created in the first instance. How are they related to previous historical conditions, past path-dependent trajectories and current circumstances? What are the sources of novelty and innovation? How is novelty and innovation created and introduced into the ongoing evolution of history? Why are some new technologies selected in preference to others?

Following this introduction, these questions will be addressed in the remainder of the paper. First, the original canonical economic explanations for path dependence are laid out as the theoretical departure point for the paper. Secondly, sociological theories of new path creation are reviewed. It is argued that these provide a theory of knowledgeable agents to explain the initial creation of new technologies within the historical context in which they are conceived. Thirdly, a hybrid socio-economic theory is proposed that seeks to explain the processes of new technological path creation in conditions of path dependence. Fourthly, some empirical evidence of new path creation is analysed. This draws on the creation of the new renewable energy technological pathway in wind power in Denmark to illustrate the theoretical arguments developed in the paper.

A final section draws these arguments and the historical evidence together. It evaluates the validity of the theoretical arguments as illustrated by the history of new path creation in the Danish wind power industry.

The Economics of Path Dependence

The seminal economic work of David (1985, 1986) on "QWERTYnomics" provides the initial inspiration for studies of technological path dependence. According to the David model, new technological pathways start as a result of small "historical accidents", "chance events" or "random" actions. Subsequently, one or more of these chance events is contingently selected for reasons not immediately connected with the original event. When that happens path dependence occurs as the original accidental events become progressively "locked-in" as the development pathway through the operation of various autocatalytic "network externalities" (David, 1985, 1986) or "increasing returns effects" (Arthur, 1989, 1994). The final feature of this canonical model is that, once lock-in has occurred it is assumed that a technology, industry, institution or industrial location pattern will persist until such time as it is disrupted by an "external shock".

This standard model was recently re-stated by Vergne and Durand (2010). They define path dependence narrowly as "a property of a stochastic process which obtains under two conditions (contingency and self-reinforcement) and causes lock-in in the absence of exogenous shock" (Vergne & Durand, 2010, p. 737). The main characteristics of this canonical model are summarized in Figure 1.

The original David model was developed to challenge two central assumptions in neoclassical, equilibrium based micro-economic theory. These are, first, that market forces will ensure that the most efficient technological solutions will finally prevail. Second, that decisions are, in principle, reversible and will be reversed if and when better technologies become available. David's (1985) example of the QWERTY keyboard and Arthur's (1990, 1992) analysis of the adoption of VHS in preference to Betamax as the standard VCR format challenged both of these assumptions.

In the first place, they argued that inefficient or sub-optimal technologies can become locked-in as industry standards and may persist for considerable periods of time. This is despite the fact that alternative and more efficient technologies are available. Thus, lock-in is one of the core concepts in the path dependence model. It expresses the idea that historical contingency combined with the emergence of self-reinforcing effects drives a technology or industry along one path rather than another and that this pathway may be based on inefficient or sub-optimal technologies, institutional or organizational arrangements.

Secondly, in place of the possible reversibility of decision-making, a key defining characteristic of path-dependent processes is "non-ergodicity". This is an inability to shake free of their history. From this perspective, path-dependent processes arrive at outcomes that evolve as a consequence of the process's own history. This history cannot be reversed. As a result, early chance events and decisions can reverberate through history closing off alternative paths and confirming a particular path with the implication that outcomes need not be based on the most efficient technologies available.

David (2005) has argued that the main point of path dependence is to restore the importance of causal, historical economic explanation involving sequential actions. This laudable aim, however, currently lacks historical explanation of how and why new technological pathways are created in the first instance. Martin and Sunley (2006) rightly

Initial conditions	**Path creation processes**	**Path establishment processes**	**Path dependence outcome**	**Path dissolution**
Not specified.	Historical accidents. Chance events. Serendipity →	Contingent selection. Self-reinforcing effects. →	Lock-in of inefficient or sub-optimal technologies, institutions or organisational forms. - - →	External shocks.

Figure 1. Basic economic model of path dependence (after David, 1985, 1986; Arthur, 1989, 1994; Vergne & Durand, 2010)
Note: Arrows indicate uni-directional historical development of pathway towards lock-in.

point out that "placing too much emphasis on random accidents as the sources of path creation constrains and undermines such causal explanation" (p. 428). Despite this much of the path dependence, literature claims that the emergence of novelty and the consequential creation of new technological pathways is essentially a serendipitous process.

The inadequate discussion of path creation in path dependence gives rise to a curious paradox in the theory. On the one hand, it is argued that historical processes are important in shaping the development of a new technology or industry once it has emerged, while on the other hand history plays no part in shaping its initial creation. But, as Witt (2003) points out new technologies do not emerge in "virgin markets". Perez (2010) also argues that "Studies of innovation (the emergence of novelty) have shown that the introduction of technical change is not random but (itself) path dependent and interdependent with other innovations clustered in systems ..." (p. 199). Thus, in reality, new technologies often have to emerge into a complex landscape of historical path-dependent developments which themselves may possibly provide significant barriers to that emergence. New path creation in such historical contexts is not explained by recourse to the notion of serendipity as suggested in the basic model of David (1985, 1986), Arthur (1989, 1994) and Vergne and Durand (2010).

The Sociology of New Path Creation

The lack of an adequate economic explanation of how new pathways are created in the first instance has led a number of scholars to turn to sociological approaches that focus on the activities of reflexive agents in new path creation. Prominent among them are Garud and Karnøe (2001). They offer an alternative theory to economic versions of path dependence. They adopt a sociological ontology and argue that any theory of new path creation should attach a significant role to the importance of knowledgeable agents and the considered "mindful deviation" of entrepreneurs from established paths. They suggest that entrepreneurs of various kinds create new pathways as they navigate the current flow of events in "real time" and seek to set new processes in motion by "mindful deviation" (Garud & Karnøe, 2001, p. 2).

The early work of Garud and Karnøe (2001) takes its inspiration from the "Social construction of technology systems" literature (e.g. Bijker *et al.*, 1987). This suggests that human agency is distributed across a multiplicity of actors who are embedded in networks and emerging technological pathways. Thus, actors both contribute to the creation and emergence of new pathways and are embedded in them as a result of their involvements in their previous historical development. But at any point in real time, these pathways provide the platforms for potential departures from them. This is summed up in the phrase that "agency associated with technological entrepreneurship is distributed, embedded, and can vary by paths" (Garud & Karnøe, 2001, p. 281).

In later work, Garud *et al.* (2010) take issue with the first four stages of the standard path dependence theory shown in Figure 1. They argue that, with respect to path creation, "initial conditions are not given but rather constructed by actors who mobilize specific sets of events from the past in pursuit of their initiatives ... what is exogenous and what is endogenous is not given but depends on how actors draw and redraw their boundaries. Emergent situations are not "contingencies" but instead afford embedded actors the possibilities to pursue certain courses of action while making others more difficult to pursue. Self-reinforcing mechanisms do not just exist but are cultivated. Rather than lock-in there

Initial conditions	Path creation process	Path establishment process	Path dependence process	Path dissolution
Constructed by actors who mobilize events from the past.	Mindful deviation by embedded actors.	Emergent contexts for action. Self-reinforcing mechanisms are cultivated.	Lock-in as a temporary stabilization of paths in the making.	Continual possibilities for creative destruction.

Figure 2. The sociology of new path creation (after Garud & Karnøe 2001; Garud *et al.*, 2010)
Note: Arrows indicate direction of historical development of new pathway and iterative reflexive feedback loops.

is ever the possibility of creative destruction … with those with the most to lose proactively making their own creations in order to survive" (Garud *et al.*, 2010, p. 769).

It is evident from these arguments that what is being proposed here is not simply a more detailed reworking of the early stages of path dependence theory, as outlined in Figure 1, but an alternative theory based on a major ontological shift to theorizing the emergence and development of a technology in terms of the interactions between actors embedded in networks. The main elements of this alternative are summarized in Figure 2.

Garud *et al.* (2010) state that they "do not embrace a path dependence perspective" (p. 762). Nevertheless, the "Social construction of technology systems" literature in some interpretations sees path dependence and path creation as continual inter-linked processes. This position is put succinctly by Martin and Sunley (2006) with respect to economic evolution which they argue may "be understood as an ongoing, never ending interplay of path dependence, path creation and path destruction that occurs as actors in different arenas reproduce, mindfully deviate from, and transform existing socio-economic-technological structures, socio-economic practices and development paths" (p. 408). The theoretical aim of this paper is therefore to develop this insight by linking evolutionary theories of path dependence with a sociological explanation of how new technological pathways are created by knowledgeable agents in the first instance in a hybrid socio-economic theory of new path creation. Before developing this new hybrid theory, it is necessary to point out that since the original narrow economic explanation of path dependence, based on contingency and self-reinforcing processes leading inevitably to lock-in that can only be broken by external shocks, additional causes of path dependence have been identified. The paper now turns to an evaluation of these arguments because they form the bases of our understanding of the evolution of the historical conditions into which new technologies emerge, and the contemporary path-dependent barriers that may possibly block their development and diffusion to a critical mass that would mark the creation of a new technological pathway.

Beyond Contingency and Self-reinforcement

The standard economic model of path dependence provided by David (1985, 1986), Arthur (1989, 1994) and Vergne and Durand (2010) is designed to show how contingency and self-reinforcing process can lead to the *outcome* of stability and lock-in of (inefficient) technologies. In this argument, the introduction of new technologies is blocked for considerable periods of time until being replaced as a result of some major external shock. But other theorists have identified a number of additional forces that can lead to path-dependent outcomes. These historical forces influence strongly the evolution of the initial conditions in which new technologies are created and the resulting contemporary barriers to new path creation.

The forces that may lead to the evolution of path-dependent development pathways include cognitive parameters in the form of technological paradigms, existing rules and norms exemplified by institutional hysteresis and established technological regimes. These barriers are all formed and maintained by path-dependent processes and are in addition to the economic forces leading to path-dependent development.

Cognitive barriers to new path creation were first explored by Dosi (1982). He developed the Kuhnian concept of scientific paradigms (Kuhn, 1962) and argued that technologies develop and diffuse along trajectories that are path-dependent because of the influence of the technological paradigms within which they are developed. Such paradigms are the prevailing models for the solution of techno-economic problems (Markard & Truffer, 2006). They may be defined as "a collectively shared logic at the convergence of technological potential, relative costs, market acceptance and functional coherence" (Perez, 2010, p. 186).

For Dosi, a technological paradigm involves a specific set of heuristics and visions on how to do things and how to improve them. They represent the conventional wisdom of the relevant community of practioners and their collectively shared cognitive frames (Dosi & Grazzi, 2010). The development pathways of new technologies proceed along trajectories that are shaped and constrained by the paradigms within which they have been inaugurated. This is a path-dependent process because the rate and direction of technical change are limited by the currently acceptable cognitive frameworks within any given technological paradigm.

Geels (2002) argues that the effects of technological paradigms are not only limited to the search heuristics adopted by engineers that strongly influence the direction of travel along a technological trajectory but also become embodied in the socio-technical landscape. This represents "a set of deep structural trends ... (expressed) in the material context of society for example the material and spatial arrangements of cities, factories, highways and electricity infrastructure" (Geels, 2002, p. 1260). In the face of such powerful cognitive and material barriers, the creation of new technical search heuristics may take years because it proceeds in small incremental steps and often involves experiments and failures (Geels, 2004).

The historical and cumulative development of knowledge along a particular technological trajectory can turn a once competitive technology into the basis of a declining industry. In this instance, learning that once functioned as "barriers to entry" for outsiders may be transformed into "barriers to exit" for insiders (Bianchi, 1986).

In a second important development of the canonical path dependence model North (1990) and Setterfield (1993, 1995, 1997) argue that institutions, in the form of socially constructed rules of behaviour, can also become locked-in. In such cases, they may provide barriers to the creation of new technological pathways. North (1990), for example, argues that the increasing returns discussed by Arthur that generate economic path-dependent development, also apply in institutions. This is because institutions develop as a web of interdependent formal and informal rules that work together to favour and reinforce some kinds of actions and exclude others.

Setterfield (1993, 1995, 1997) also developed the idea that institutions can become barriers to change. He introduced the concept of institutional hysteresis in which institutions and economic pathways are argued to co-evolve in an interdependent way. In the short run, institutions are said to be "exogenous" to the economy and to provide a stable framework for economic activity. This initial stability may not be conducive to the emergence of new technologies. As path dependence sets in, however, it is also argued that institutional structures become "endogenous" to the economy and open to feedback effects from changes in the economy. This leads to the co-evolution of path-dependent institutions and economic pathways. Echoing the canonical theory of path dependence North and Setterfield both argue that, as with technologies, some institutional structures that emerge may not be the most efficient. As a result of their path-dependent development, this means that some inefficient and sub-optimal institutions may become locked-in for long periods of time (Martin & Sunley, 2006). Thus, in conditions of institutional hysteresis where inefficient and sub-optimal economic pathways and institutions co-evolve together and become locked-in to these forms for considerable periods of time, they can form hostile historical conditions and powerful contemporary barriers to new technological path creation.

Technological regimes provide a third type of barrier to new path creation. This concept was first introduced by Nelson and Winter (1977). As part of their development of an evolutionary approach to economic change, they argued that firm behaviour is governed by a set of technological routines. The combined set of such routines they defined as a technological regime. They noted that the problem-solving activities of engineers were not oriented so much to changes in costs or demand conditions, as predicted by neoclassical micro-economic theory, but remained remarkably stable over time. They were focused on particular problems and informed by certain ideas of how these problems might be solved. Nelson and Winter (1977) illustrate these ideas using the long-term and incremental development of the DC3 aircraft programme during the 1930s.

This original, narrow definition of a technological regime was broadened by the sociologists Rip and Kemp (1998) to incorporate much more of the social context of technology. In contrast to Nelson and Winter's (1977) narrow definition, they defined a technological regime as "the whole complex of scientific knowledges, engineering practices, production process technologies, product characteristics, skills and procedures, and institutions and infrastructures that make up the totality of a technology" (Rip & Kemp, 1998, p. 340). In this conception, technological regimes represent an intermediate level between specific innovations as they are conceived and developed by particular agents and what Rip and Kemp (1998) call "social landscapes". They argue that technological regimes represent on the one hand the boundary conditions within which technologies are developed and on the other links to the more concrete and slow to change social landscape.

The idea that there are different degrees of technological path dependence at different analytical levels has been influential in subsequent studies of the relationships between individual technologies and whole technological systems and their stability and change. Geels (2004), for example, takes up the sociological approach of Rip and Kemp (1998) to analyse the barriers to the adoption of new technological systems. He analyses transitions from one technological system to another in terms of three path-dependent analytical dimensions. Technological regimes together with other types of regime form the meso-level of this analysis. Above them in a hierarchy of relative stability and hardness lies the socio-technical landscape that includes the material aspects of society and is therefore difficult and slow to change. Below the meso-regime level, a multiplicity of technological niches come and go.

This idea of multiple levels of technological path dependence may also apply to institutional path dependence described above. Thus, at the landscape level, institutions may take the form of legal rules enacted by governments. At the grass roots level, they may be the historical ways of doing things. And at the meso-level, they may be informal but conventional rules that incur social sanctions if they are not followed.

Overall, the processes of new path creation are confronted by three major types of path-dependent barriers in addition to basic economic barriers such as how competitiveness and costs are defined, risk aversion and network externalities. These barriers arise because of the processes leading to path dependence and different degrees of lock-in. They include cognitive barriers in the form of technological paradigms, institutional hysteresis and social barriers emanating from the groups who are the bearers of existing technological and other types of regimes.

Hybrid Theory

Given the historical evolution of path-dependent development trajectories that form the bases of the initial conditions and barriers, outlined above, that confront the creation of new technological pathways, the evolutionary hybrid theory proposed in this paper seeks to explain how new technological pathways are created in conditions of historical path-dependent development. The theory starts with the assumption that new pathways are not created by disembodied economic forces but by knowledgeable agents. In order to explain the reflexive actions of these agents, it is necessary to add a sociological dimension to the analyses of the causes of path dependence.

In their sociological alternative to traditional economic path dependence theory, Garud and Karnøe (2001) and Garud *et al.* (2010) argue that path dependence is a *process* that is influenced by history but that is also made out of that history as a result of the interactions between embedded agents and the structures in which they are embedded. In this case, technological deviations and departures may be introduced, mostly incrementally, over time by agents of change. In their earlier work, Garud and Karnøe (2001) argue that the key agents are entrepreneurs who mindfully deviate from contemporary flows of events to create new pathways. Later they broaden this view to include "also those who develop complementary assets ... those in institutional forums ... and customers who offer critical inputs that shape emerging paths" (Garud & Karnøe, 2003, p. 279). This list, however, does not exhaust the potential actors who may contribute to the creation of new technological knowledge and economic novelty which may be regarded as the key drivers of new path creation.

Upstream from entrepreneurs who may be responsible for the commercialization of new knowledge in the form of innovations, individual and collective inventors play key roles. In the past, individual private sector inventors have been the pioneers of new technologies, for example, in successive Kondratieff waves of technological development. But, the increasing scientific basis of new technologies has also been accompanied by a growing role for large R&D intensive companies (e.g. pharmaceuticals) and public sector actors such as universities (e.g. genetic technologies) and government research establishments (e.g. nuclear technologies). Unlike entrepreneurs selling into competitive markets, organizations such as these are often supplying markets for such services as health and power that have been established or regulated by governments. Thus, a free market model in which private sector entrepreneurs are the main agents of new technological path creation has limitations in its potential applicability in contemporary Western economies, particularly those characterized by "co-ordinated capitalism".

Garud and Karnøe (2003) also ask what mindfully deviating actors actually do to create new pathways. They suggest two possible strategies and illustrate them with an analysis of the early path creation of wind power technologies and industries in Denmark and the US. They argue that the early pioneers in Denmark adopted a strategy of "bricolage". This is defined as "a process of moving ahead on the basis of inputs from actors who possess local knowledge, but through their interactions, are able to gradually transform emerging paths to higher degrees of functionality" (Garud & Karnøe, 2003, p. 296). In other words, actors in this case sought to improve the existing technologies by incremental innovations. This created a new wind power pathway by modest adaptations over a long period of time.

In contrast, actors in the NASA-led US public wind power research programme adopted a strategy of "breakthrough". This is defined as "actors attempting to generate dramatic outcomes rather than adaptiveness, an unyielding vision to leap-frog the Danish initiative" (Garud & Karnøe, 2003, p. 279). In this case, the actors involved were trying to generate new technologies by radical innovations. In the event, the Danish actors' strategy succeeded while that of the US actors was mostly an expensive failure.

There is also a substantial literature from other fields such as historical sociology and political science on the activities of knowledgeable agents that may lead to the creation of new institutional arrangements and new technological pathways. Of particular note here is the introduction to their book by Streek and Thelen (2005). In it they identify three ways in which the "gradual transformation" (p. 9) of institutional arrangements may take place. These are displacement, layering and conversion. Such discontinuities and transformations may equally apply to the actions of knowledgeable agents in new path creation. They represent three empirically substantiated ways in which the sociological interactions of actors may "transform emerging paths to higher degrees of functionality" (Garud & Karnøe, 2003, p. 296) and thereby create new technological pathways. Such transformation may also involve purposive planning (Sydow *et al.*, 2010).

Displacement is defined as the "slowly rising salience of subordinate relative to dominant institutions" (Streek & Thelen, 2005, p. 31) and, in the argument advanced here, technologies. Defection from one to the other is the mechanism by which change takes place. This is made possible by the existence of some incoherence within existing arrangements that opens up possibilities for "mindful deviation", the cultivation of alternative technological paradigms, the rediscovery of dormant technologies and/or the assimilation of external practices.

The early history of the motor car provides one example of this process at work. During the late nineteenth century, the previous long-term development of steam power led to this emerging as an important early propulsion technology for the new "horseless carriage". Even Henry Ford spent many of his early years trying to build a steam road carriage and a farm locomotive. But, after he moved to Detroit, he learned that a German engineer, Nicholas Otto, had built an internal combustion engine driven by petrol. From then on, Ford concentrated on developing a petrol-driven motor car. Once he went into mass production with his Model T, steam was no longer developed as a serious propulsion technology for motor cars.

Layering involves adding new rules, procedures, structures or technologies to what already exists. Geels (2002) provides an example of this process at work with his analysis of the rise of steam ships. He argues that, initially, steam engines were added to sailing ships as an auxiliary device. In this way, the new technological pathway was created by adding the new form of propulsion to what already existed. The subsequent differential growth of steam technology eventually led to the dominance of this new technology over the old.

Conversion is the reorientation of an institution or technology in terms of its form or function. An example of this process at work is that, in the early days of radio its main function was to facilitate the communication of information concerning phenomena such as the weather and maritime movements. It was not until the 1920s that its main function began to shift and convert towards broadcast entertainment. This created the new production pathway for the production of content for this new function of the medium.

Taken together these different types of agent activities suggest that transformative new technological pathways can be created in ways other than those involving external shocks, "gales of creative destruction" and terminal path breaking. These are relatively rare events. Breakthrough innovations and their rapid diffusion to create new pathways are also the exception rather than the rule.

The hybrid theory advanced in this paper is therefore primarily concerned with explaining new path creation by knowledgeable agents in the context of transformations of the path-dependent characteristics of the economic environment, technological paradigms, institutions and technological regimes. These provide powerful historical barriers to change. Rather than confronting them directly, it is often easier for agents to initiate new technologies in niches that are not subject to the full force of the tides of history. Niches may be regarded as important minority selection environments in which the process of new path creation may be started. "They provide locations for learning processes, e.g. about technical specifications, user preferences, public policies, symbolic meanings. Niches are locations where it is possible to deviate from the rules of the existing regime" (Geels, 2004, p. 912). While not all new technologies start in such niches, they often provide important potential locations for the process of mindful deviation that is required for new path creation (Garud & Karnøe, 2001).

A "niche" may be defined as an application context in which the new product or technology is temporarily protected from the standards and selection rules of the prevailing paradigm (Kemp *et al.*, 1998; Hoogma *et al.*, 2002; Markard & Truffer, 2006). Niches provide space for novelties to incubate without being subjected to prevailing competitive market pressures or the normal selection criteria that accompany the dominant technological trajectories. The informal rules of niche environments are less articulated and subject to higher degrees of uncertainty than those of the established trajectories (Geels, 2004). In niche conditions, it is also possible to draw on new or deviant local or international knowledge in order to develop new business networks, value chains and user–producer relationships.

While niche environments provide opportunities for technological inventions to be introduced, they do not of themselves create inventions. Those are introduced as a result of interactions between knowledgeable economic actors, policy-makers and the potential users of inventions or their results. Following Garud and Karnøe (2001), it is argued that strategic agency is significant in new path creation. This involves the "mindful deviation" from established paths not only of individual private sector entrepreneurs but also, among others, of researchers in universities, large companies and government research establishments.

The commercialization of inventions and the creation of new pathways in the form of innovations is more often than not a long and incremental process. Frequently, older pathways continue for some time while newer ones are initially introduced as additions to existing technologies rather than immediate replacements for them. The nature of these processes is captured by the concepts of "displacement", "layering" and "conversion" (Streek & Thelen, 2005).

Niches may shelter initially technological inventions from the path-dependent tides of the economic selection environment, technological paradigms, institutions and technological regimes. Nevertheless, eventually these barriers have to be overcome if they are to be able to diffuse to a critical mass and the creation of a new pathway in the overall technological landscape.

A critical mass is reached where, in the face of existing barriers, sufficient economic agents are prepared to switch to the new alternatives (Witt, 1997). Critical mass is a well-known phenomenon in non-linear dynamic systems (Lorenz, 1993). It may be defined as a point of discontinuity that induces a dramatic turn away from an existing system (Witt, 1997). The current economic selection environments, technological paradigms, institutions and regimes always favour the existing and widely used product variants. For this reason, success in the creation of new technological pathways comes down to the prerequisite to pass a critical mass (Witt, 1997).

The diffusion of innovations is a key process that spreads their use from their original invention in niche conditions to the point at which critical mass is achieved and a new economic pathway is created that represents a significant discontinuity in the technological landscape. In contemporary economies, as with the introduction of an innovation, diffusion also requires knowledgeable agents. The roles of diffusion agents have a long pedigree in the diffusion literature (Brown, 1981; Rogers, 1995). Thus, when a major innovation is introduced, commercial marketing agencies or in-house people are often given the task of promoting them and triggering a diffusion process (Witt, 1997).

Innovation diffusion may also need to be organized by collective action. This is because setting the diffusion process in motion is like providing a public good. Without collective action, the early adopters would have to bear the initial network diseconomies while later adopters would profit from the investments of the early adopters (Witt, 1997). Thus, public action may be needed to overcome some of the barriers to new path creation. The introduction of electric cars is one example where such action is necessary.

The theoretical arguments proposed in this paper are summarized in a simplified form in Figure 3. For the purposes of exposition, the figure is divided into five main elements. In the first, it is argued that the initial conditions confronting the introduction of a new technology are determined by previous rounds of the historical evolution of path-dependent technological development trajectories. Given these initial conditions, the new path creation process starts with the mindful deviation of knowledgeable agents. In the case of new technologies, these agents are defined as "inventors". Such deviation is easier in niches that are not subject to the full force of the initial conditions. In a second stage of the new path creation process, inventions are taken up by innovators who introduce transformations and discontinuities into the contemporary mix of technologies. In order to reach critical mass and to become a new technological pathway, these introductions

Initial conditions	**Path creation process**	**New path establishment processes**	**Barriers to new path creation**	**Landscape change outcome**
Historical evolution of path dependent technological development trajectories.	Mindful deviation and invention by knowledgeable agents in niches.	Incremental innovation by knowledgeable agents by displacement, layering or conversion. Or radical breakthrough innovation.	Historical economic selection environment. Cognitive: technological paradigms. Institutional hysteresis. Social: technological & other regimes.	New technology diffuses to achieve critical mass and tipping points. Or de-locking of existing pathways. Or continuation of path dependent development trajectories.

Figure 3. Hybrid socio-economic theory of new path creation
Note: Arrows indicate direction of trajectory of new path creation and reflexive feedback loops.

have to overcome one or more of four main path-dependent barriers to new path creation. These are the economic selection environment, the relevant existing technological paradigms, institutional hysteresis and the relevant contemporary technological regimes. The outcome of the tensions and conflicts between innovation and the barriers to new path creation may be either the diffusion of a new technology to critical mass and the creation of a new pathway, the de-locking of an existing pathway or the continued lock-in of path-dependent development trajectories. It is not possible to predict in advance which of these three outcomes will prevail in the economic landscape.

This hybrid approach to understanding new path creation is illustrated in this paper by the introduction of new wind power technologies in Denmark following WWII. This new pathway was created in the context of the established path-dependent technological trajectories for the generation and supply of electricity.

This hybrid is an evolutionary theory in the terms proposed by Martin and Sunley (2006). It seeks to establish an empirically grounded basis for understanding new technological path creation in the context of the "ongoing, never ending interplay of path dependence, path creation and path destruction that occurs as actors in different arenas reproduce, mindfully deviate from, and transform existing socio-economic-technological structures, socio-economic practices and development paths" (Martin and Sunley, 2006, p. 408).

New Path Creation: The Introduction of Wind Power Technologies in Denmark

Initial Conditions

At the beginning of the twentieth century, the established technological pathways for the generation and supply of electricity did not apply much beyond urban areas. This provided significant niches in rural areas that were not connected to grid supplied AC electricity. As a result, inventions and innovations in the generation of electricity using wind power were first developed and used in rural areas. There they were used to produce direct current (DC) electricity powering batteries. This type of current cannot be transmitted over long distances via a grid system.

After the War restrictions were imposed on Danish diesel fuel imports. The supply of diesel for rural electricity generators was uncertain. This added to the niche conditions for the adoption of wind generators in rural areas.

Confidence in the reliability and price of oil supplies was also shaken by the external shock of the Yom Kippur War in 1973 and the Iran/Iraq war of 1980–1988. They sparked widespread political debates. These led to a switch to coal-based energy generation and the emergence of a strong anti-nuclear movement that rejected nuclear power as a long-term solution to Denmark's energy needs.

Path Creation Processes

The first Danish wind turbines were invented in the rural niche beyond the contemporary reach of urban based grid connected electricity suppliers. The most influential pioneering inventor was Poul la Cour. He wanted to bring the benefits of electricity to people living in the countryside. In order to achieve this, he managed to mobilize resources from the Danish government to build his first DC windmill in Askov east of Esbjerg in 1891. This was a smock type mill with a fantail to point the sails into the wind and self-reefing shuttered sails. In the same year, he also invented a device called a "kratostat" to smooth out the power fluctuations caused by variable wind speeds.

In Denmark, early innovations in AC wind turbines were inspired by deviation from the dominant technological paradigm and the development of an alternative in a local minority culture of enthusiasm for wind power. Johannes Juul was one of the key pioneer inventors. He was a student of Poul la Cour and managed to mobilize resources from the Danish utility SEAS to fund a niche research and development programme to produce electricity using wind power. As a result of this programme, the first Danish AC wind turbine was built in 1950 at Vester Egesborg on the south-west coast of Zealand. It was grid-connected using an induction generator and was left in service for a decade. The successful demonstration of Juul's experimental turbines enabled him mobilize further resources by securing public niche funding from the Danish government to construct a larger model in 1958 at Gedser about 130 km south of Copenhagen.

These early incremental inventions gradually increased the power output of wind turbines. But none of them could generate more than 200 kW of electricity. At this level of productivity, they were not competitive in the markets outside their various niches. It became clear that power outputs in excess of 1 MW would be needed to make wind turbines economically competitive with other sources of energy at "normal" prices.

The first reliable 1 MW wind turbine was constructed by the Teacher Group in Tvind. Like Juul, they started by deviating from the dominant fossil fuel-based technological paradigm for the generation of electricity. They did not accept that nuclear power was an acceptable alternative to this paradigm. As a result, they set about demonstrating that wind power could provide electricity on a scale and at a price acceptable to electricity utility companies. They were led by pioneer innovator Amdi Peterson. To keep costs down, they purchased a second-hand generator and gearbox originally used on a mine hoist. A low-speed shaft was bought second-hand from a shipyard. Volunteers built the concrete tower. They went to Stuttgart to learn from Ulrich Hutter how to make variable pitch fibreglass blades. The grid was supplied via a rectifier converter that was purpose built for the turbine by students from the technical University of Denmark supervised by Ulrik Krabbe. The wind turbine was completed in 1978 and remains in operation in 2010. It laid the foundation by which Danish wind turbines have diffused to become commercially competitive in world markets.

In 1979, the Danish parliament passed legislation that extended the niche market for wind turbines by providing wind turbine purchasers with a 30% subsidy as long as the turbine was of an approved type. In addition, in the late 1970s, the connection of small AC turbines to the national grid in Denmark was formalized by Danske Elvaerkers Forening (Danish Association of Electricity Supply Undertakings). This established a collective ability for anyone with an approved wind turbine to connect to the national grid and to receive payments for any surplus electricity they supplied to the network. These niche market conditions were further extended with legislation that allowed individuals who had shares in co-operatively owned turbines in or near where they lived, to pay no tax on the income from electricity sales up to about 9000 kWh. This together with the 30% capital subsidy made well-sited and maintained wind turbines an attractive financial investment for local individuals and co-operatives.

Path Establishment Processes

The diffusion of wind-powered DC generators developed slowly between the wars. During this period, the manufacturing company Lykkegaard was an important diffusion agent. They made and installed around 20 machines before the Second World War. As a result, the Lykkegaard Company made and installed around 60 of their four-bladed windmills in the early 1940s. Even so, despite strong niches and weak barriers to creation of this new technological pathway, the total number of installed wind generators in Denmark remained small at this time.

The diffusion of the pioneer inventions by la Cour and Juul to create a new industrial pathway in Denmark required a combination of both niche home market users and knowledgeable manufacturing diffusion agents. The main users were individual farmers and local rural co-operatives. The main diffusion agents were agricultural machinery manufacturers. The incremental innovations made by the early pioneers and subsequently by wind enthusiasts demonstrated the practicality of generating grid-connected AC electricity using wind turbines. Danish agricultural machinery manufacturers saw a market opportunity to diversify their product ranges. Their long experience of engineering robust and reliable agricultural equipment made Danish wind turbines as reliable as was possible for an entirely new product.

One of the first diffusion agents to diversify in this way was Vestas. In the 1970s, it was a leading Danish manufacturer of mobile cranes and agricultural equipment employing about 100 people. It diversified by buying the manufacturing rights to the successful

HVK wind turbine developed by Karl Erik Jorgensen and Henrik Stiesdal. Production started in 1979. The firm is now the largest producer of wind turbines in the world with over 13,000 employees in 2007 (Musgrove, 2010, p. 107).

Path-dependent Barriers to New Path Creation

Although Denmark was the first country in which a significant new technological pathway was created in wind turbines, this was achieved in the face of a number of path-dependent barriers. The first of these was the economic hostility of existing electricity generators. They had long-term investments in fossil fuel-based generation technologies. Their arguments regarding the relative competitiveness of these technologies, when compared with renewable alternatives, were also framed in such a way as to exclude their consequential costs such as pollution and waste disposal. As a result, the diffusion of wind turbine technology was slowed until its purely financial costs could converge with those of conventional technologies. This could only be achieved by the development of reliable 1 MW and larger turbines.

The diffusion of wind turbine technology also had to overcome the prevailing technological paradigm of centrally generated, AC, grid-connected electricity generation. The conventional wisdom of this paradigm was that matching the supply with the demand for electricity required reliable base load generation, some flexible capacity for meeting peaks in demand and technological systems for matching both peaks and troughs in demand with the available supply. In the context of this paradigm, wind power was seen as an intermittent source of energy not able to supply either base load requirements or peak demand. It was also seen as requiring additional technological systems in order to make use of its energy in the first instance and to accommodate both local consumers' exports and imports of electricity at different times. Overcoming these barriers required the development of complementary technological systems that could cope with the intermittent supply of energy and institutional changes requiring existing electricity generators to make some use of wind energy.

The diffusion of wind turbine technology was also confronted not so much by institutional hysteresis as by institutional churn. At some points in time such as 1985, 1990 and 1996, the Danish Government specified that Danish energy companies should make use of wind power. But at others such as 1996 and 1999 this was placed in doubt as a result of the EU seeking to open up national electricity markets to international suppliers over which the Danish government had little control. Thus, the development of the new pathway did not follow a continuous sequence of positive stages. There were also setbacks in its development. Such institutional churn brought into conflict the long-term investment strategies required to introduce the new technology with short-term political and institutional changes. Overcoming this barrier required long-term certainty in the market for renewable electricity in order to stimulate the investment required to diffuse widely the new technology.

Landscape Change

By 1980, the new wind turbine manufacturing pathway had been created in Denmark. At that time, there were a dozen manufacturers producing grid-connected AC wind turbines with rotors ranging from 5 to 15 m producing 5–55 kW of electricity. Further incremental innovations increased their efficiency throughout the 1980s. This increase in productivity was sufficient to compensate for the removal of the Danish government's 30% capital purchase subsidy in 1989.

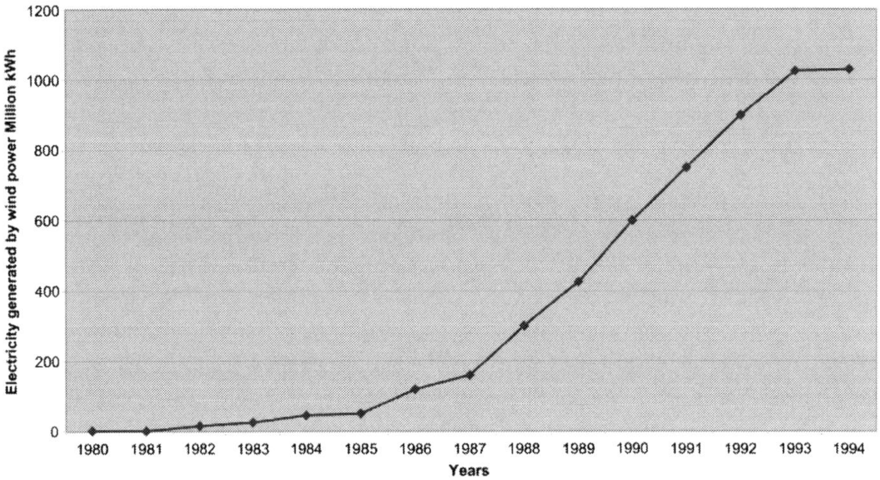

Figure 4. Denmark: new path creation, electricity generation by wind power 1980–1994
Source: Adapted from Gipe (1997).

During the 1980s and 1990s, this new pathway developed to supply some 6.65% of Danish electricity by 2004. Figure 4 shows the take-off and growth of the contribution of the new technological pathway to the supply of electricity in Denmark. The graph illustrates the argument in this paper that the Danish inventions and innovations in wind energy technologies led to the creation of a new indigenous economic pathway that makes a significant new contribution to the supply of electricity in Denmark.

The Danish wind turbine pathway was created over a long period of time starting with the inventions of Poul la Cour at the end of the nineteenth century and culminating with the successful diffusion of the technology by firms such as Vestas, Bonus, Micon and Nordtank in the 1980s. The initial conditions in which the new technology developed were set by previous rounds of path-dependent development. In the early days, these were the limited reach of electricity grid systems in rural areas. Later on, they also included the initial competitive market advantages of fossil fuel-based centrally generated electricity. It was started by the mindful deviation of knowledgeable actors in rural and publicly created niches. It was developed primarily through incremental innovations that added a new layer of electricity generation technology to the existing fossil fuel-based technologies.

This layering process involved adding new rules in the form of legalizing grid connection of approved wind turbines; procedures in terms of government subsidies, tax relief and feed in tariffs; structures in the form of government regulations and technologies in the form of wind turbines and complementary technological systems to the existing electricity supply system in Denmark.

The early pioneers were inspired by alternative technological paradigms either in terms of the supply of electricity in rural areas or by an anti-nuclear animus. It affected and was affected by the existing technological landscape for the generation of electricity. Thus, on the one hand, it added a new competitive technology to that generation landscape and on the other its diffusion was made easier by oil price shocks to the conventional fossil fuel-based electricity supply landscape. These iterative interactions are summarized in Figure 5.

Initial conditions	**New path creation process**	**New path establishment processes**	**Barriers to new path creation**	**Landscape change**
Limited conventional energy resources. Limited reach of grid connected electricity supplies. Uncertainty of oil imports after WW II. Yom Kippur War 1973 & Iran/Iraq war 1980-88 generate oil price shocks. Switch to coal based electricity generation. Decline of agricultural sector. Social movement critical of conventional energy sources, rejecting nuclear alternative OOA & OVE.	SEAS funds R&D for Juul to build first AC turbine 1950 followed by Govt. funded turbine 1958 at Gedser. 1975 teacher Group pioneer reliable 1MW turbine at Tvind. Late 1970s DEF formalises grid connection of small turbines. 1979 Govt. introduces 30% subsidies for turbines approved by Risø National Laboratory. Tax relief for local individual owners of wind turbines & co-operatives. 1984 DWTOA sets feed in tariffs at 85% of consumer prices.	Inter-war & 1940-43 Lykkegaard Co. manufactures & installs 80 turbines. Interactive learning from bottom up. 1977 Grove-Nielsen starts ØVE fibreglass blades. 1979 Vestas crane & agricultural machinery co. buys IPR to Jorgensen & Stiesdel HVK turbine. 1980s manufacturing & diffusion by Vestas, Bonus, Micon & Nordtank Cos. Exporting 70-80% of production. 1985, 1990 Govt. specifies that utilities install 100 MW & 1996 200 MW of wind generation capacity.	*Economic* Hostility of existing utilities. Assessment of viability that excludes consequential costs of pollution or waste disposal. *Cognitive* Prevailing fossil fuel based tech. paradigm. *Institutional* 1996 EU directive on liberalisation of electricity markets combined with 1999 Danish Electricity supply Act. Reversion with change of govt. in 2001.	Creation of new pathway: 1980 dozen manufacturers producing AC grid connected turbines. 2008 indigenous industry employs 28,000 workers. Development of small-scale capitalism – 80% of 6,300 turbines in 2002 owned by individual farmers or co-operatives. 6.65% of Danish electricity produced by wind power in 2004.

Figure 5. New technological pathway created in wind power (Denmark 1950–1980s)

Conclusions

It has been argued in this paper that the canonical model of path dependence does not attempt to provide an explanation of the processes associated with the intentional creation of new technological pathways. It is primarily concerned with the emergence of path dependence. Nevertheless, new pathways are not created by disembodied economic forces. They require social action by knowledgeable pioneering individuals, universities, companies and/or governments. The aim of the hybrid theory proposed in this paper is to provide a potential empirically grounded basis for understanding the interactions between the evolution of path-dependent economic development and the social actions of pioneering agents who contribute to new path creation.

In this context, it is argued that new path creation is an iterative process in which the initial conditions are set by previous rounds of path-dependent development. In these conditions, knowledgeable agents who are identified as inventors and innovators may set in motion the creation of new technological pathways by consciously deviating from past practices and introducing and attempting to diffuse new technologies. The general ways in which new technologies are diffused by agents include the adoption of incremental strategies such as displacement where the salience of subordinate technologies increases to overtake that of the dominant technology, layering, where new technologies are added to those already present, and conversion where old technologies are changed.

These strategies do not require that existing pathways are completely broken in order to facilitate the creation of new ones. In the case of displacement, a dominant technology is superseded by a subordinate one but may continue for some time afterwards. Layering involves the addition of a new technology while previous technologies also continue to be employed. Conversion makes use of previous knowledge to develop a new technology. This is the closest to a path-breaking scenario, although remnants of the old may continue in the new.

Given that the canonical version of path dependence argues that lock-in is the outcome of the process and that new pathways are created in initial conditions of previously locked-in development pathways, path-dependent processes themselves establish major barriers to new path creation. Foremost among these barriers are cognitive lock-ins in the form of existing technological paradigms, institutional lock-ins in the form of institutional hysteresis and technological regime lock-ins in the form of the entrenched power of the social groups who are the bearers and principle protagonists of those regimes.

The outcomes of the iterations between knowledgeable agents and forces involved in the different stages of new path creation are unpredictable. They may lead to changes in the existing technological landscape as a new technology diffuses to a critical mass at which point it could be described as constituting a new pathway. Alternatively, the barriers to diffusion may be so great as to prevent new path creation and maintain previous path-dependent development trajectories.

A final element in this iterative model is that the technological landscape itself can be either a force for continuing lock-in or a force for new path creation. Contemporary concerns about climate change, for example, if translated into effective collective action can change dramatically the technological landscape in which power is generated and used.

These arguments have been illustrated by case studies of the initial introduction of wind power technologies for the generation of electricity in Denmark. This is only one example and further historical case studies and empirical analyses would be necessary to demonstrate the degree of general applicability of the theoretical approach outlined in this paper.

The Danish case study suggests that new technological pathways are created initially in niches. In the case of wind power, these consisted of rural areas where the dominant urban, centralized and grid-connected electricity generation and supply system did not exist or were markets created by government fiscal policies that protected the new technology from "normal" market competition. New niches were also created by changes at the level of the overall technological landscape. They included rapid changes in the relative competitiveness of fossil fuels as a result of war and revolution in the Middle East and climate change.

The knowledgeable agents of new path creation included individual inventors and innovators. Sometimes, their inventions were not taken up at all and at others it required considerable periods of time for them to be developed into innovations. The diffusion agents for these innovations were either private production companies or developers exploiting niche conditions.

The creation of the new technological pathway in wind power was not marked by radical breakthrough innovations. Indeed, in the US, for example, government attempts to sponsor radical innovation by contracting different types of firms and organizations to attempt to leapfrog existing levels of technological development proved unsuccessful. Instead, new pathways were created by incremental innovations that either added the new technology to existing ways of generating electricity or recombined older technologies in new ways.

The path-dependent barriers to new path creation were formidable. In some cases, they were consciously challenged by innovators. This was the case where the prevailing technological paradigm was deliberately rejected either to bring electricity to rural areas or to demonstrate a viable alternative to nuclear power. In other instances, the power of the social groups supporting existing technological regimes was weakened by landscape developments such as rapid rises in oil prices, climate change and nuclear disasters.

As a result of these interactions, the technological landscape for electricity generation has changed slowly over a period of half a century or more. The process of new path creation and the use of wind power have been largely incremental. Nevertheless, the EU as a whole now has a target of 20% of energy generation to be provided by renewables by 2020. The majority of this will be provided by wind power. Thus, this source of electricity has now diffused to the point at which it can be described as a new technological pathway.

References

Arthur, W. B. D. (1989) Competing technologies, increasing returns, and lock-in by historical events, *Economic Journal*, 99 (394), pp. 116–131.

Arthur, W. B. D. (1990) Positive feedbacks in the economy, *Scientific American*, 262 (February), pp. 92–99.

Arthur, W. B. D. (1992) Learning and adaptation in the economy. Santa Fe Institute Working Paper 92-07-038, Stanford University, Stanford.

Arthur, W. B. D. (1994) Path dependence, self-reinforcement and human learning, in: W. B. D. Arthur (Ed.) *Increasing Returns and Path Dependence in the Economy*, pp. 133–158 (Michigan, MI: Michigan University Press).

Arthur, W. B., Durlauf, S. N. & Lane, David A. (Eds) (1997) *The Economy as a Complex Evolving System II* (Reading, MA: Perseus Books).

Bianchi, P. (1986) Industrial restructuring within an Italian perspective. Working Paper No. 2, Nomisha, September.

Bijker, W. E., Hughes, T. P. & Pinch, T. J. (1987) *The Social Construction of Technological Systems* (Cambridge, MA: MIT Press).

Brown, L. A. (1981) *Innovations Diffusion—A New Perspective* (London: Methuen).

David, P. A. (1985) Clio and the economics of QWERTY, *American Economic Review*, 75(2), pp. 332–337.

David, P. A. (1986) Understanding the economics of QWERTY, in: W. N. Parket (Ed.) *Economic History and the Modern Economist*, pp. 30–49 (Oxford: Blackwell).

David, P. A. (Ed.) (2005) *Path Dependence in Economic Processes: Implications for Policy Analysis in Dynamical Systems Contexts. The Evolutionary Foundations of Economics* (Cambridge: University Press).

Dopfer, K. (Ed.) (2004) *The Evolutionary Foundations of Economics* (Cambridge: Cambridge University Press).

Dosi, G. (1982) Technological paradigms and technological trajectories: A suggested interpretation, *Research Policy*, 11(3), pp. 147–162.

Dosi, G. & Grazzi, M. (2010) On the nature of technological knowledge, procedures, artefacts and production inputs, *Cambridge Journal of Economics*, 34(1), pp. 173–184.

Dosi, G., Freeman, C., Nelson, R., Silverberg, G. & Soete, L. (Eds) (1988) *Technical Change and Economic Theory* (London: Frances Pinter).

Fagerberg, J. (2003) Schumpeter and the revival of evolutionary economics: An appraisal, *Journal of Evolutionary Economics*, 13, pp. 125–159.

Foster, J. (1997) The analytical foundations of evolutionary economics: From biological analogy to economic self-organisation, *Structural Change and Economic Dynamics*, 8(4), pp. 427–451.

Garud, R. & Karnøe, P. (2001) *Path Dependence and Creation* (London: Lawrence Erlbaum Associates).

Garud, R. & Karnøe, P. (2003) Bricolage versus breakthrough: Distributed and embedded agency in technology entrepreneurship, *Research Policy*, 32, pp. 277–300.

Garud, R., Kumaraswamy, A. & Karnøe, P. (2010) Path dependency or path creation? *Journal of Management Studies*, 47(4), pp. 760–774.

Geels, F. W. (2002) Technological transitions as evolutionary reconfiguration processes: A multi-level perspective and a case study, *Research Policy*, 31, pp. 1257–1274.

Geels, F. W. (2004) From sectoral systems of innovation to socio-technical systems: Insights about dynamics and change from sociology and institutional theory, *Research Policy*, 33, pp. 897–920.

Gipe, P. (1997) *Wind Energy Comes of Age* (New York, NY: John Wiley).

Hodgson, G. (1993) *Economics and Evolution. Bringing Life Back into Economics* (Oxford: Polity Press).

Hoogma, R., Kemp, R., Schot, J. & Truffer, B. (2002) *Experimenting for Sustainable Transport. The Approach of Strategic Niche Management* (London: Spon).

Kemp, R., Schot, J. & Hoogma, R. (1998) Regime shifts to sustainability through processes of niche formation: The approach of strategic niche management, *Technology Analysis & Strategic Management*, 10(2), pp. 175–195.

Kuhn, T. (1962) *The Structure of Scientific Revolutions* (Chicago, IL: University of Chicago Press).

Lorenz, H. W. (1993) *Nonlinear Dynamical Economics and Chaotic Motion* (Berlin: Springer-Verlag).

Markard, J. & Truffer, B. (2006) Innovation processes in large technical systems: Market liberalization as a driver for radical change? *Research Policy*, 35, pp. 609–625.

Martin, R. & Sunley, P. (2006) Path dependence and regional economic evolution, *Journal of Economic Geography*, 6(4), pp. 395–437.

Metcalfe, S. (1998) *Evolutionary Economics and Creative Destruction* (London: Routledge).

Metcalfe, J. S. & Foster, J. (Eds) (2004) *Evolution and Economic Complexity* (Cheltenham: Edward Elgar).

Musgrove, P. (2010) *Wind Power* (Cambridge: Cambridge Uuniversity Press).

Nelson, R. R. & Winter, S. G. (1977) In search of a useful theory of innovation, *Research Policy*, 6(1), pp. 36–76.

Nelson, R. R. & Winter, S. G. (1982) *An Evolutionary Theory of Economic Change* (Cambridge: Belknap Press).

North, D. C. (1990) *Institutions, Institutional Change and Economic Performance* (Cambridge: Cambridge University Press).

Perez, C. (2010) Technological revolutions and technological paradigms, *Cambridge Journal of Economics*, 34(1), pp. 185–202.

Potts, J. (2000) *The New Evolutionary Micro-economics: Complexity, Competence and Adaptive Behaviour* (Cheltenham: Edward Elgar).

Rip, A. & Kemp, R. (1998) Technological change, in: S. Rayner & E. L. Malone (Eds) *Human Choice and Climate Change,* vol. 2, pp. 327–399 (Columbus, OH: Battelle Press).

Rogers, E. M. (Ed.) (1995) *Diffusion of Innovations* (New York, NY: Free Press).

Schumpeter, J. A. (1939) *Business Cycles: A Theoretical, Historical and Statistical Analysis of the Capitalist Process* (New York, NY: McGraw-Hill).

Schumpeter, J. A. (1942) *Capitalism, Socialism and Democracy* (New York, NY: McGraw-Hill).

Setterfield, M. (1993) A model of institutional hysteresis, *Journal of Economic Issues*, 27(3), pp. 755–774.

Setterfield, M. (1995) Historical time and economic theory, *Review of Political Economy*, 7(1), pp. 1–27.

Setterfield, M. (1997) *Rapid Growth and Relative Decline: Modelling Macroeconomic Dynamics with Hysteresis* (London: Macmillan).

Streek, W. & Thelen, K. (2005) *Beyond Continuity: Institutional Change in Advanced Political Economies* (Oxford: Oxford University Press).

Sydow, J., Lerch, F. & Staber, U. (2010) Planning for path dependence? The case of a network in the Berlin–Brandenburgh optics cluster, *Economic Geography*, 86(2), pp. 173–195.

Vergne, J.-P. & Durand, R. (2010) The missing link between the empirics of path dependence: Conceptual clarification, testability issue, and methodological implications, *Journal of Management Studies*, 47(4), pp. 736–759.

Witt, U. (Ed.) (1993) *Evolutionary Economics. The International Library of Critical Writings in Economics* (London: Edward Elgar).

Witt, U. (1997) "Lock-in vs. critical masses—industrial change under network externalities", *International Journal of Industrial Organization*, 15(6), pp. 753–773.

Witt, U. (2003) *The Evolving Economy* (Cheltenham: Edward Elgar).

Comparing Alternative Path Creation Frameworks in the Context of Emerging Biofuel Fields in the Netherlands, Sweden and Finland

RAIMO LOVIO* & PAULA KIVIMAA**

*School of Economics, Aalto University, Finland, **Finnish Environment Institute, Helsinki, Finland

Abstract *Several studies on innovation have brought forward different conceptions on how innovations and new technological paths are created. Theoretical arguments differ regarding the foundational ontologies and the major factors influencing path formation. The article aims to compare two strands of literature related to new path formation: (1) strategic niche management and the related multi-level perspective and (2) technological innovation systems. The article examines how these models operate in the context of empirical analyses hypothesizing that there may be fewer differences in empirical results than in theoretical argumentation and that new empirical findings can give a good impetus also for theoretical elaborations. Thus, the article compares empirical analyses made in the same context, namely the transport biofuel field. The actual development of biofuel fields in the Netherlands, Sweden and Finland are also compared. Our analysis suggests some areas in which the two theoretical frameworks might be developed to take recent empirical observations and the spatial dimension better into account.*

1. Introduction

Technological innovations for energy production and use have been studied from multiple perspectives, such as new path creation (Garud & Karnøe, 2001), technological innovation systems (TISs) (Jacobsson & Bergek, 2004; Hekkert *et al.*, 2007), strategic niche management (SNM) (Schot & Geels, 2008) and the related multi-level perspective (MLP) (Geels, 2002) and energy entrepreneurship (Wüstenhagen & Wuebker, 2011). Whether the different approaches lead to similar or contradictory conclusions and recommendations is a key question from the point of view of supporting innovation in practice.

This article makes a multi-paradigm analytical effort by examining how different "path creation" frameworks work at the level of empirical analysis. It looks at two elaborated theoretical frameworks used to study new energy technology paths: TISs and SNM and the related MLP. Some comparisons of these approaches have previously been made by Geels *et al.* (2008) and Markard and Truffer (2008). Instead of theoretical foundations and ontologies of the approaches (Garud *et al.*, 2009; Geels, 2009, 2010a), this article mostly focuses on the empirical approaches and results of the frameworks regarding clean energy technology.

Focusing merely on theoretical comparison might exaggerate the differences regarding the driving forces of clean energy technologies. Our hypothesis is that the differences of the approaches regarding their empirical results may be smaller than those of their theoretical foundations. In addition, we are trying to find new insights or complementary cross-overs between the two approaches.

TIS and SNM/MLP were chosen, because they share a certain degree of policy orientation. This is important for comparing policy recommendations arising from the empirical analyses. In addition, we can find studies examining the same development (transport biofuels) and the same localities (The Netherlands and Sweden) under the two approaches. For comparing similarities and differences, the same empirical context is of immense value.

As a secondary objective, the article compares the role of different factors on the actual development of biofuel fields in the Netherlands, Sweden and Finland. The comparison of these countries is based on a list of categorizations that arise from the comparison of TIS and SNM. New empirical data are added for the Finnish case. Our analysis suggests some areas in which the two theoretical frameworks might be developed to take empirical observations better into account.

In Section 2, TIS and SNM/MLP are briefly introduced and compared at a general level. Section 3 compares the empirical analyses broadly, including the description and use of methods and empirical data. Section 4 focuses on comparing biofuel analyses and their conclusions. Section 5 takes the country approach, and the results are discussed and concluded in Section 6.

2. Introduction to TIS, SNM and MLP

The two frameworks examined have been much discussed. Therefore, the following introduction focuses on the analytical core elements. As the frameworks continually evolve and different scholars may highlight different aspects of the approaches, the frameworks are mainly presented as used in the context of biofuel analyses.

2.1 *Technological Innovation Systems*

The TIS approach was originally developed in Sweden (Jacobsson & Johnson, 2000) and later used and further developed in the Netherlands (e.g. Hekkert *et al.*, 2007; Hekkert & Negro, 2009; Suurs, 2009). It has been used for analysing the emergence of new energy technologies in countries such as Sweden, Germany and the Netherlands. A TIS has been described as a socio-technical system focused on the development, diffusion and use of particular technologies such as wind turbines or biofuels. TISs may have a geographical dimension, but are often international in nature. Yet most analyses based on the TIS approach have been made at a national level.

According to the "Swedish version", an analysis based on the TIS approach consists of three basic steps (Bergek *et al.*, 2008a). The starting point for the analysis is defining the TIS and identifying its structural components: actors, networks and institutions (Carlsson & Stankiewicz, 1991). Later, another structural component, namely technology, including both artefacts and knowledge, was added to the list (Bergek *et al.*, 2008b). Describing the structural components provides a basis for the second step: analysing the TIS in functional process terms. The functions, or key processes, constitute an intermediate level between the structural components of the TIS and their performance. The list of functions used in the TIS approach has varied during the years. Bergek *et al.* (2008b) list eight functions including knowledge development, entrepreneurial experimentation, materialization, influence of the direction of search, market formation, resource mobilization, legitimation and the development of positive externalities. In the empirical analysis, the extent to which key processes or functions are fulfilled is analysed by using various indicators. The interactions between functions and the functional pattern that dominates the development of the TIS are essential.

The third step in the analysis is to explain the identified level and form of the functional pattern of the TIS. This is done by Bergek *et al.* (2008a) in terms of inducement and blocking mechanisms. These mechanisms explain why some key processes work well, whereas others do not. The mechanisms that shape the functional pattern of the TIS can be internal or external to the TIS. An internal blocking mechanism may, for instance, be poorly developed learning and political networks that hinder knowledge diffusion and legitimation of the new technology. An external blocking mechanism may come in the form of highly organized incumbents that defend their markets and investments (Jacobsson, 2008).

Dutch scholars have further developed the TIS approach. The following summary of their arguments is mainly based on a recent analysis of the formative stage of natural gas as an automotive fuel (Suurs *et al.*, 2010). In their description of the structural components, three basic categories are distinguished: actors, institutions and technologies. In an actual TIS, the components are linked to each other and, if they form configurations, they are called networks. Thus, the structural components are specified in a similar manner as in the above-described "Swedish version".

Related to the second step, the process or function level, a number of elaborations have been made by the Dutch scholars. First, the list of system functions is different (Table 1). Second, the functional pattern of the TIS is called motors of innovation. Dutch scholars have distinguished at least three motors of innovation. The entrepreneurial motor is dominated by entrepreneurial activities function. The system building motor is dominated by conscious orientation of almost all actors on building the TIS. The market motor is a pattern in which the promise of real market development is the primary driver of event sequences (Hekkert *et al.*, 2007). Third, Suurs *et al.* (2010) enhance the analysis of major inducement and blocking mechanisms by relating them to the structural components and to the historical pattern of cumulative causation of the TIS (motors of change). According to this analysis, major institutional drivers are often external to TIS (such as oil scarcity). This illustrates the fact that external factors play an important role and, therefore, have to be included in the analysis, while the endogenous dynamics of the system can be explained by system functions and their interaction.

Table 1. Functions of TISs (Suurs *et al.*, 2010, p. 421)

System function	Description	Event types associated
Entrepreneurial activities	Entrepreneurs are at the core of any innovation system. These risk takers exploit business opportunities and perform innovative commercial and/or practice-oriented experiments	Projects with a commercial aim, demonstrations, portfolio expansions
Knowledge development	Non-technological knowledge and technological R&D are a source of variation in the system and, therefore, prerequisites for innovation processes to occur	Studies, laboratory trials, pilots
Knowledge diffusion	The typical organizational structure of an emergent innovation system is the knowledge network, primarily facilitating information exchange.	Conferences, workshops, alliances
Guidance of the search	Selection processes are necessary to facilitate a convergence in development	Expectations, promises, policy targets, standards, research outcomes
Market formation	New technologies often cannot outperform established ones. In order to stimulate innovation, it is necessary to facilitate the creation of (niche) markets, where new technologies have a possibility to grow	Market regulations, tax exemptions
Resource mobilization	Financial, material and human factors are necessary inputs for all innovation system developments	Subsidies, investments
Support from advocacy coalititions	The emergence of a new technology often leads to resistance from established actors. For an innovation system to develop, actors need to raise a political lobby that counteracts inertia and supports new technology	Lobbies, advice

2.2 *SNM and MLP*

SNM and MLP have been developed, mainly in the Netherlands (e.g. Kemp *et al.*, 1998; Geels, 2002). They are presented here mainly as described by Schot and Geels (2008).

SNM scholars argue that, for many innovations, market niches and user demand are not readily available, because these innovations are not minor variations on the prevailing set of technologies but differ radically. At the start, a technological niche provides a protected space that nurtures a specific set of interactions between actors and issues. Such niches function as proto-markets that may jumpstart the development of market niches, involving entrepreneurs, users, policy-makers and societal groups who invest their time and resources to enhance the niche. Early SNM work conceptualized a bottom-up process, in which novelties emerge in technological niches, then conquer market niches and eventually replace and transform dominant technologies and regimes (Schot & Geels, 2008).

Based on insights from various sources, in SNM three internal processes were distinguished for successful development of a technological niche: the articulation of expectations and visions, the building of social networks and learning processes in multiple dimensions. Subsequently, more specific hypotheses were formulated for each process (Table 2).

Niche innovations are rarely able to bring about great transformations without the help of broader forces and processes. This conclusion led to the MLP that linked niche internal and external processes by distinguishing three analytical levels: niche, regime and landscape. Niches form the micro-level where radical novelties emerge. The socio-technical regime forms the meso-level, which accounts for the stability of existing large-scale systems. The macro-level is formed by the socio-technical landscape, an exogenous environment beyond the direct influence of niche and regime actors. The core notion of the MLP is that transitions come about through interactions between processes at different levels: (a) niche innovations build up internal momentum, (b) changes at the landscape level create pressure on the regime and (c) destabilization of the regime creates windows of opportunity for niche innovations (Schot & Geels, 2008).

Since the basic MLP model was introduced (Geels, 2002), further developments have taken place. These can be summarized into four major findings. First, the emerging and successful niche innovations need not always compete with and substitute the prevailing regimes; they may also be incorporated into existing regimes (hybridization solutions) (Geels & Schot, 2007). Second, niche developments may be influenced by multiple regimes and niche innovations interact. This broader attention to multiple regimes and multiple niches may generate new modifications in transition pathways (Geels, 2007;

Table 2. Niche internal processes and related hypotheses (based on Schot & Geels, 2008; van der Laak *et al.*, 2007)

Internal processes	Description	Hypotheses for successful niche building
The articulation of expectations and visions	Crucial for niche development because of providing direction to learning processes, attracting attention, and legitimating protection and nurturing	Expectations are shared by many actors, they are specific enough to give guidance, and their content is substantiated by ongoing projects
The building of social networks	Creates a constituency behind the new technology, facilitates interactions between relevant stakeholders and provides necessary resources (money, people, expertise)	Networks are broad (multiple kinds of stakeholders are included) and deep (participants are able to mobilize commitment and resources within their own organizations) and alignment within the network is facilitated through regular interactions between the actors
Learning processes at multiple dimensions	Technical aspects, market and user preferences, cultural meaning, infrastructure, production networks, regulations, and societal and environmental effects	Learning processes include first-order learning (accumulation of facts and data) and reflexive second-order learning (enabling changes in frames and assumptions) and are broad focusing on technological, economic and social aspects

Raven, 2007). Third, the destabilization of old regimes is one explanation to niche success (Geels & Turnheim, 2010). Fourth, innovation and transition processes include also socio-cognitive factors, politics and power play, the depth of which is easily underplayed if a narrow technological perspective is adopted (Geels, 2010b).

2.3 *Comparison at a General Level*

Geels *et al.* (2008) see a lot in common in the two approaches when compared with many other (such as neo-liberal, ecological modernization, deep ecology). For example, in both approaches, the primary unit of analysis is the organizational field. In both TIS and MLP, socio-technical transitions are seen as sustainable innovation journeys to capture the open and uncertain nature of radical technological changes, to highlight the agency dimension and to refrain from simple policy recipes such as "getting the prices right". In either approach, cultural studies and discourse analyses have not been prominent before the most recent analyses.

Yet, in other respects, the approaches differ. First, the major disciplinary background of SNM/MLP is in sociology (science and technology studies) and evolutionary economics, whereas the roots of TIS are additionally in industrial economics and management science. The domination of traditional quantitative methods in industrial economics and management science might lead to TIS scholars being more interested in using quantitative data and measurements than SNM/MLP scholars.

Second, the approaches differ in how they build explanations to the success or failure of new path creation. While TIS and SNM/MLP both make a distinction in terms of "levels", MLP has progressed further in conceptualizing the interactions between internal and external processes and between different levels (niche, regime and landscape). On the other hand, TIS's functional analyses (Table 1) appear more elaborated than SNM's three niche internal processes (Table 2) (see also Markard & Truffer, 2008), although SNM gives more attention to less tangible processes such as creating expectations and learning. In empirical analyses, these differences may realize as a clearer search for societal explanations by SNM/MLP scholars, while the TIS scholars focus on the internal functions of new technological systems.

Third, regarding actors, the TIS scholars emphasize the role of entrepreneurs (system function 1, Table 1), whereas SNM argues more strongly for the importance of broad stakeholder networks (Table 2). This difference may lead the TIS scholars to focus their empirical analyses and explanations on actions by incumbents and new firms.

According to our hypothesis (Section 1), however, the above differences in theoretical and methodological choices may prove to be smaller in actual empirical analyses and findings. This will be studied in Sections 3 and 4.

Geels *et al.* (2008) draw several common robust policy lessons from empirical analyses made by using these approaches. Similar findings have also been found in many other innovation studies (Kivimaa, 2008). The TIS and SNM/MLP findings related to biofuel analyses are later compared with this list of policy lessons.

- For radical innovations, new markets co-develop with new technologies. This view deviates from the linear model, which sees markets materialize after a technology is fully developed.
- Consistent and stable policy frameworks are important for innovation journeys.

- Social embedding of technology and socio-political legitimacy are important, for both early market formation and further up-scaling.
- Interactions between multiple innovation journeys can generate positive external economies that flow across technological innovations systems.
- The role of incumbents and new entrants varies in different cases and contexts.
- Policy-makers should have sufficient competence and substantive expertise of the technologies, actors and networks in order to avoid one-sided policy measures, for example, due to uncritical accepting of demands or promises of strong actors.
- Policies need to be multifaceted and designed to target many functions and internal processes.

3. Methodological Issues

Comparison of the two theoretical frameworks involved two stages. First TIS and SNM/MPL articles related to transport and energy sectors were searched to find comparable empirical studies. Then five articles using similar empirical cases were used for further analysis. The empirics of the five articles on Sweden and the Netherlands were complemented with Finnish data to gain further insight into the role of geographical and spatial diversity in path creation (Kivimaa & Mickwitz, 2011; Lovio *et al.*, 2011b). Our empirical analyses have been guided by a flexible utilization of the theoretical ideas and empirical methods used by both TIS and SNM/MLP scholars and other path creation/dependence literature (see Lovio *et al.*, 2011a).

Table 3 includes the selection of 14 recent empirical analyses made by using the TIS, SNM and MLP approaches. The comparison reveals that many commonalities can be found in the research designs, such as the emergence of one new energy technology (or two) in one or two European national contexts. In addition, most of the articles examine emergence processes after the oil crises of the 1970s and the analyses cover a time period between one to three decades. The empirical research designs based on MLP differ from the other articles: time periods are much longer, and more importantly, the focus is not only on the emergence of a new technology or TIS but also on the destabilization of an old technology or the transition from an old socio-technical system to a new one.

Typical for most articles in Table 3 is that the space devoted to explaining research data and analytical methods is rather limited. The authors seem to be more ambitious in the area of theory than that of empirical data and methods. In more recent articles, however, there are two features in the empirical analyses that evidence some changes. First, the differences of the analyses to variable-based approaches and explanations are underlined. Second, a shift towards placing more emphasis on socio-cognitive and socio-political aspects is visible.

With respect to both MLP and in the "Dutch version" of the TIS approach, it is explicitly argued that these approaches are "process theories" (Suurs & Hekkert, 2009; Geels & Turnheim, 2010). Process theory focuses not on independent variables but on events that are the results of actors who react to each other and make decisions on the basis of beliefs and interests that change over time. The MLP or SNM scholars have not developed or used any specific new methods in analysing event chains. Their analyses have been based on traditional qualitative and historical reasoning aimed at providing analytical

Table 3. A selection of recent articles based on TIS, SNM and MLP with a focus on technologies related to energy and transport sectors

TIS approach "Swedish version"	TIS approach "Dutch version"	SNM	MLP
Jacobsson and Bergek (2004): wind turbines, solar cells and collectors in the Netherlands, Germany and Sweden 1990–2001	Hillman *et al.* (2008): biofuels in the Netherlands and Sweden 1980–2005	van der Laak *et al.* (2007): biofuels in the Netherlands 1990–2005	Geels (2005): from horse-drawn carriages to automobiles in 1860–1930
Jacobsson and Lauber (2006): wind turbines and solar cells in Germany 1974–2003.	Suurs and Hekkert (2009): biofuels in the Netherlands 1990–2007	Ulmanen *et al.* (2009): biofuels in the Netherlands and Sweden 1973–2007	van Bree *et al.* (2010): hydrogen and battery-electric vehicles in the future
Jacobsson (2008): biopower in Sweden 1980–2006	Suurs *et al.* (2010): natural gas as an automotive fuel in the Netherlands 1970–2007	Sengers *et al.* (2010): biofuels in the Netherlands 2000–2008	Geels and Turnheim, (2010): the coal industry in the UK 1913–1970
Bergek (2010): wind power planning in Sweden		Raven and Geels (2010): biogas development in Denmark and the Netherlands 1973–2004	

histories oriented towards finding patterns and mechanisms. Instead, the Dutch TIS scholars have applied event history analysis developed by Van de Ven and colleagues (Van de Ven *et al.*, 1999; Poole *et al.*, 2000) as a quantitative empirical tool when identifying the occurrence and timing of functions. In an empirical analysis by Suurs and Hekkert (2009), events were identified by using digital databases covering a large number of various sources and then mapped on appropriate system functions. They argue that this method allows a fruitful combination of quantitative and qualitative analysis but they do not argue that quantitative methods and "objective data" could replace "the interpretation of the researcher as a crucial factor" (Suurs & Hekkert, 2009, p. 1006).

Geels *et al.* (2008) discovered in their comparative analysis that both TIS and SNM/MLP could learn from cultural studies and policy analysis, placing more emphasis on socio-cognitive and socio-political processes. In recent articles shown in Table 3, this shift in empirical research is visible. Ulmanen *et al.* (2009) state that they utilized two types of conceptual lenses, SNM and the political science concepts of "advocacy coalitions" and "discourses". Later, Sengers *et al.* (2010) used content analysis on a sample of biofuels articles to deepen the understanding of social acceptance of biofuels in different discourses. Bergek (2010) also used content analysis to study wind power planning and permitting procedures. The shift from techno-economic and institutional "facts" towards socio-cognitive interpretations and socio-political interests can be seen as a response towards criticisms that these approaches have encountered (e.g. Meadowcroft, 2009). It can also be seen as a reaction to empirical findings. The emergence of wind power and biofuel paths has not only been resisted by established regime actors, as both approaches have predicted, but sometimes also by local citizens and environmental movements.

In conclusion, the differences between the empirical methods used appear smaller than could be expected based on the background disciplines of TIS and SNM/MLP. The shared opinion seems to be that research questions and phenomena are decisive when deciding on the empirical methods.

4. Are There Differences in Key Findings of the SNM and TIS Biofuel Analyses?

By focusing the comparison of empirical analyses made by using TIS and SNM/MLP on five articles focused on the same empirical subject, biofuel development in the Netherlands and Sweden, this section examines the similarities and differences in terms of major explanations, actors and policy advice between Dutch and Swedish transport biofuel paths.

Biofuels include two distinct technology groups: first-generation (1G) and second-generation (2G) biofuels. 1G is based on conventional technologies and mainly agricultural crops, such as rapeseed, wheat or sugar beet, to produce biodiesel or bioethanol. The 2G biofuels originate from chemical and biotechnological industries involving the conversion of woody biomass, waste wood or cultivated energy crops, into "biocrude", thus not directly competing with food production. For 2G biofuels, the distance-to-market is much larger than for 1G biofuels, but they offer possibilities for larger reductions in CO_2 emissions (Suurs & Hekkert, 2009). 1G fuels can only be used as a relatively small component in ordinary vehicles or as high-concentration mixes in special flexifuel vehicles (FFVs).

Biofuels contributed less than a percent to total European transport fuel consumption in 2003. In 2003, EU Directive 2003/30/EC set a 5.75% biofuels target for 2010 and later the target was increased to 10% by 2020. Sweden was among the few EU countries able to reach the 2010 target due to the largest FFV fleet in Europe. Recently, the dramatic shift in the debate on biofuels—from a highly promising technology to leading to more harm than good—has complicated reaching the 2020 target. Anti-biofuel lobbies include not only activists but actors from environmental science arguing against the use of specific feedstocks (e.g. palm oil) or more generally demanding strict criteria for sustainable production of biofuels.

It is commonly found that the biofuel path started to emergence earlier and is stronger in Sweden than in the Netherlands (Table 4). The TIS and SNM/MLP scholars explain the difference in quite a similar way, although their approaches include variation in explanatory structures as noted in Section 2.3. Both scholars attribute the difference to more consistent and multifaceted policy in Sweden that reduces long-term uncertainty, plays down the potential controversy between 1G and 2G biofuels and balances the voices of pro-biofuel and anti-biofuel coalitions, at least so far. When elaborating their explanations, SNM/MLP scholars Ulmanen *et al.* (2009, pp. 1413–1415) even refer to some TIS articles.

The TIS and SNM findings are also similar in that policy-induced niche markets often precede more extensive system developments. While they use different explanatory factors, many similar elements have been fitted into both frameworks with the exception that resource mobilization has not explicitly emerged as a central element of SNM-based

Table 4. Key elements between biofuel path creation in the Netherlands, Sweden and Finland (based on sources in Table 3 and Lovio *et al.*, 2011b; Kivimaa & Mickwitz, 2011)

	The Netherlands	Sweden	Finland
Nature and status of biofuel path	Started in the early 1990s, lack of continuity, today's slow development mainly EU policy-driven. Temporary niche markets, no solid niches prior to EU Biofuels Directive. Major turning points: the Biofuels Directive and increasing oil prices	Started in the 1970s, high continuity, good interaction between experiments, policies and corporate actions at local, national and EU levels. A few viable market niches. Major turning points: 1970s oil crises, local policies and the Biofuels Directive	Started as a reaction to the EU Biofuels Directive in 2003. Some research activities prior to that. Major turning points: in the late 2000s ambitious targets set exceeding EU requirements
Actors and networks	Networks comprising agricultural organizations, regional authorities, public transport companies, and SMEs. Some networks involved research organizations, government agencies and incumbent companies. Local entrepreneurs and authorities the most active actors	Agriculture, forest and energy sector cooperation supported by the car industry. Strong R&D networks involving public energy agency, universities, forest sector research organizations or farmers' cooperatives. Local entrepreneurs and large national companies	First, mainly forest industry-related R&D networks. Later national oil companies as the key actors, forest companies as potential feedstock suppliers. Local entrepreneurs and experiments rare
Institutions and market formation	Temporary case-specific tax exemptions and regional policy support combined with tightening water quality legislation as a motivator for trials. National government inactive, mainly focused on technology push. EU Biofuels Directive setting a distribution obligation	Strong public R&D support and tax exemptions since the 1970–1980s. From 2000, further tax reductions and local incentives, e.g. free parking. Joint activities between government agencies to support market formation/diffusion, successful introduction of FFVs, municipal bus companies using biofuels	National R&D support. Market formation supported later by the distribution obligation and changes in fuel taxation
Competing niches	Controversy between 1G (local experiments) and 2G (national policy), ethanol and biodiesel	1G as a bridge to 2G. Controversy between agricultural crops and wood, methanol and ethanol	Focus on 2G only

(Continued)

Table 4. Continued

	The Netherlands	Sweden	Finland
National resources	Agricultural crops	Forest-based resources and agricultural crops	Forest-based resources and limited agricultural crops
Existing regime influence	Decline of the agricultural sector versus the existence of established fossil fuel regime	Car and forest industries' R&D activities and the decline of the agricultural sector	Lack of forest industry's interest in transport biofuels before the Biofuels Directive
Spatial issues	EU influence combined with regional activities, resulting in regional niche markets	National policy support for R&D and regional trials. Sweden active to lobby supporting policies in the EU and exceed EU requirements	EU policy initiating national policies and business activities
Current discourses and advocacy coalitions	Controversial views on the benefits of biofuels, i.e. environmental and agricultural discourses	Discourses based on oil substitution, environment and agriculture with high emphasis on the oil discourse	Competing pro-biofuel and anti-biofuel advocacy coalitions

studies and learning has been less explicit in TIS-studies. TIS identifies the role of public policy through at least three system functions. In the Dutch biofuel case, regional governments had a guiding role (guidance of search), increasingly stringent local water quality regulation triggered a small niche market (market formation) and resources were also mobilized by public authorities (resource mobilization) (Hillman *et al.*, 2008). Simultaneously, policies were also described among exogenous factors. In SNM, policies related to all process categories. In the Dutch biofuel case, more stringent water quality regulation shaped expectations together with temporary tax exemptions, policy-makers participated in the building of social networks and experiences from permitting barriers constituted as learning (van der Laak *et al.*, 2007; Ulmanen *et al.*, 2009).

Interestingly, both approaches also bring forward the role of advocacy coalitions as an important factor for new path creation, e.g. in the Dutch case lobbying for tax exemptions (e.g. Hillman *et al.*, 2008; Ulmanen *et al.*, 2009). The role of path-dependent existing systems is somewhat addressed through the decline of the agricultural sector as the start for entrepreneurial and networking activities.

Regarding actors, the most likely difference between the TIS and SNM approaches was that TIS focuses more on entrepreneurial activities whereas SNM highlights policy issues, users and social acceptance of biofuels more (Section 2.3). This expectation proved to be only partly true. In addition to demonstrating the social acceptance issues, SNM/MLP scholars (Ulmanen *et al.*, 2009) refer as much as TIS scholars to company activities in Sweden and in the Netherlands, which is then used to partly explain the differences between path creation in these countries. However, the difference between the approaches shows in the presentation of final conclusions. SNM/MLP scholars tend to highlight the importance of network activities and interactions in advancing new technology. Instead, the TIS scholars highlight private and municipal entrepreneurs as prime movers (Hillman *et al.*, 2008, p. 606) following their argument that "at the core of any innovation systems are the entrepreneurs" (Table 1).

As noted earlier in relation to policy lessons, both TIS and SNM/MLP scholars argue that neo-liberal strategies focusing on "getting the prices right" are limited and insufficient to bring about the required socio-technical system changes, because sustainable transitions not only entail changes in prices and markets but also in technologies, user practices, policy and cultural discourses and governing institutions. While accepting this view, we cannot overlook the fact that in the empirical case studies both TIS and SNM scholars identify price and market issues as strong explanatory factors. First, oil price development shaped the form and timing of biofuel development and diffusion. Second, scholars substantiate the important role of tax exemptions, for example, arguing that "without tax exemptions little would have happened" (Ulmanen *et al.*, 2009, p. 1413). Third, the EU directive obliging oil companies to distribute an amount of biofuel equal to 2–10% of the total distributed transport fuel have proved to be a strong market creation measure.

If market prices, tax exemptions and direct market generating measures play such great roles, it seems unsatisfactory to keep these only as external factors or leave them out of the analysis, merely stating that "another limitation of SNM is its lack of focus on economic processes. However, many other approaches exist for developing economic policies" (van der Laak *et al.*, 2007, p. 3221). In fact, Ulmanen *et al.* (2009, p. 1413) propose a more valid response to these empirical findings. They suggest placing more emphasis on examining the ways in which various actors and advocacy coalitions succeed or fail in acquiring financial protection in different forms.

The contradictions between general policy lessons as presented in Section 2.3 and empirical observations seem to support our general argument that the differences between various approaches are smaller at the empirical level than at the theoretical level and that it could be fruitful finding complementary crossovers even between the TIS and SNM/MLP approaches, focusing on institutional, socio-political and cognitive contexts and the "neo-classical" preoccupation with prices and markets.

5. How Much do National Characteristics and Different Spatial Levels Matter?

TIS and SNM scholars both have concluded that the biofuel development in Sweden and the Netherlands varies significantly (Hillman *et al.*, 2008, p. 605; Ulmanen *et al.*, 2009, p. 1413) (Table 4). They explain this by dividing the differences between national endogenous and exogenous factors. Hillman *et al.* (2008, p. 608) argue that the differences between the Swedish and Dutch cases can "only be explained to a limited extent by exogenous factors, suggesting the importance of endogenous forces—system functions—in stimulating BIS dynamics". Thus, entrepreneurs, policy-makers and other relevant actors in Sweden have been more capable than their counterparts in the Netherlands to generate actions leading to virtuous forms of cumulative causation. On the other hand, Hillman *et al.* (2008, p. 609) weaken the argument by adding that "many differences at the macro-level between the two countries with regards to cultural, industrial and political structures and events have not been considered as possible alternative explanatory variables".

Ulmanen *et al.* (2009, p. 1413) elucidate the endogenous and exogenous factors in more detail:

> The Swedish position as a neutral country and the lack of fossil fuel resources differ from the Dutch one. The Netherlands occupy a strategic geographical location in the flows of feedstock, including oil, and moreover, the Netherlands have been a major natural gas power in Europe. This is also reflected in the dominant position of Shell in the Dutch networks and partly explains the focus on the supply of fuels. In Sweden independence in case of emergency and the amble availability of biomass resources have led to an early and continued interest in biofuels. Also, in the Swedish networks the car industry has played a much more prominent role. The supply of FFVs can be understood from this perspective . . .

Based on a comparison of the Swedish, Dutch and Finnish cases, the explanatory factors for path creation include geographical factors (e.g. location, availability of biomass resources), industrial structures and existing regimes (e.g. dominant players in the energy and agriculture sectors) and political factors (e.g. Swedish neutrality). Importantly, the studies reveal that the biofuel paths have different industrial sectors behind the development. In the Netherlands, the agricultural lobby has been strong, whereas in Sweden the development has been a joint effort between forest and car industries. Finland has been a latecomer in the transport biofuel field, despite some brief statements in early 1980s energy policy (Kivimaa & Mickwitz, 2011). Supporting attitudes and policies towards 1G biofuels were lacking, first because of the non-existence of a strong domestic agricultural overproduction lobby, and later because the scientific life cycle analyses convinced policy-makers in the early 2000s that, in the Finnish context, the production of 1G biofuels is not an economically and environmentally relevant option compared to forest-based 2G biofuels.

While in Sweden, an anticipation of the EU Directive 2003/30/EC on biofuels supported action in the transport biofuels field (Kivimaa, 2008; Ulmanen *et al.*, 2009), Finland opposed the directive and postponed its enforcement. In 2009–2010, Finnish policy dramatically changed. First, Finnish oil companies launched new technologies to produce 2G biofuels from various feedstocks including waste (Lovio *et al.*, 2011b). Second, incumbent forest companies became interested in developing technologies— investigated earlier also in Sweden and the Netherlands—to produce biodiesel from wood-based feedstock. This move was motivated by the decline of pulp and paper demand in Europe, which forced forest companies to search for new products and markets (Kivimaa & Kautto, 2010). In 2011, the three largest forest companies applied public domestic and EU investment support for building commercial wood-based biodiesel plants and the Parliament decided that Finland aims to reach a 20% biofuel share in total transport fuel consumption, exceeding the demands of the EU (Lovio *et al.*, 2011b). Thus, Finland differs from the other two countries by having a path created almost solely by policy that was responded by incumbent companies instead of a variety of local experiments in different actor-network constellations.

The comparison between countries in Table 4 shows that actors and networks, institutions and technology—the key elements shared by SNM and TIS—are not enough to explain path creation in clean energy technologies. A resources/geographical dimension need to be added. Yet even when resource-based geographical similarities are as high as they are between Finland and Sweden, other structural factors related to existing regimes (both an early lack of interest of the forest industry and an absence of car industry in Finland) and differences in policy support create major variation in emerging technology paths. Similar differences have been highlighted in other new energy technology areas. Simmie

(2012) summarizes his analysis on wind power development in Denmark and California by concluding that the locations of the production sites of these new industries were not a matter of serenpidity.

The real analytical challenge is how, in an empirical analysis, the influences of structural national characteristics are mediated through endogenous socio-cognitive micro-processes (Raven & Geels, 2010, p. 97), and how the spatial dimension in general could be included in the SNM/MLP and TIS frameworks. One of the first explorations in this respect by SNM scholars is Coenen *et al.* (2010) who discuss how the SNM approach relates to proximity advantages in innovation processes as identified in the geography of innovation literature. For the TIS approach, a natural step could be building explicit links to the regional and national innovation system literature, which also uses system concepts and to which TIS scholars often refer to (e.g. Bergek *et al.*, 2008a).

The Dutch, Swedish and Finnish biofuel cases demonstrate two other reasons also for making the spatial dimension more prominent when empirically examining innovation and transition journeys in the energy field. First, the Dutch TIS scholars convincingly point out that the biofuel development in Sweden and the Netherlands can only be understood and explained if policy support or resistance at different levels, namely the regional, the national and the EU level, and the interaction between these levels, is taken into account. In the Dutch case, the earlier entrepreneurial experiments were most strongly related to regional initiatives. Later, the development was largely the result of European pressure and related regional efforts, whereas the national government generally hampered rather than supported the development (Hillman *et al.*, 2008, p. 607; Suurs & Hekkert, 2009, p. 1017). In the Swedish case, regional and national initiatives developed more in tandem, and therefore the role of EU directives was not as crucial. In Finland, path creation was completely EU-driven before major nation-wide corporative initiatives together with EU support transformed national policy. Regional actors' role in Finland was insignificant.

An important argument for the relevance of the spatial dimension is also given by Sengers *et al.* (2010) when they map the media and practitioner discourses on biofuels, revealing that the underlying structure of the biofuel discourse can be depicted by two factorial axes "techno-economic *vs.* social-ecological" and "regional *vs.* global": "Both practitioners and media coverage with techno-economic/regional frame have more confidence in technological solutions and show a positive attitude towards biofuel applications. On the other hand, both practitioners and media coverage with a social-ecological/global frame were fierce in criticizing and resisting biofuel applications for having negative side effects with regard to the environment and people in developing countries" (Sengers *et al.*, 2010, p. 5026). Thus, regarding biofuels, regional economic benefits may compete with global environmental and social worries in people's minds.

6. Discussion and Conclusions

We have compared two influential path creation frameworks in the area of energy transitions, the TIS and the SNM/MLP approaches, focusing specifically on empirical analysis including research designs and methods (Section 3), empirical focus and findings (Section 4) and the spatial dimension (Section 5). We argued that these approaches differ theoretically mainly in three aspects (Section 2.3), while their empirical approaches and findings are not significantly different. The comparison of TIS- and SNM-based analyses on biofuel path creation supported this hypothesis, although there were differences in terminology and some in the foci of empirical analysis. First, although TIS scholars have used quantitative elements, the general character of empirical analyses based on both TIS and SNM/MLP is mainly qualitative and based on multi-source historical analyses. Second, we did not find any contradictions in the major empirical findings and explanations regarding the Dutch and Swedish biofuel paths. Third, both approaches equally described the role of entrepreneurs and business activities, and only in concluding remarks, the TIS scholars tended to put more emphasis on the role of entrepreneurs, while SNM scholars highlighted learning and knowledge in networks more.

Section 2.3 summarized seven common robust policy lessons from empirical analyses made by using these approaches. Four of them proved to be important in the case of biofuel developments in Sweden, the Netherlands and Finland. The success of the biofuel field in Sweden compared with that in the Netherlands and Finland can be explained foremost by Sweden's multifaceted and stable policy framework (Lessons 2 and 7). Also, Swedish policy-makers avoided one-sided policy measures favouring either 1G or 2G biofuels, which enhanced long-term development from the easiest option towards more radical changes (Lesson 6). Moreover, the social embedding and socio-political legitimacy of biofuels have been so far less controversial in Sweden than in the Netherlands and in Finland (Lesson 3).

Both the TIS and SNM/MLP scholars end up with the conclusion that the development of the biofuel field seems to be at a crossroad. On the one hand, there is a great pressure to replace oil with biomass and emerging biofuel markets are strongly supported by the EU. On the other hand, the problems related to biofuel feedstocks and the economic feasibility of 2G technology are not yet solved. Our data from Finland confirm this conclusion. If the technical and economic problems related to 2G biofuels can be solved in the near future, biofuels can play a role in decreasing the CO_2 emissions of transport in the EU. In addition, many other developments, such as the increasing use of smaller and more efficient cars and of electric vehicles, the development of public transport and reduction in transport demand, need to take place to achieve the CO_2 emission targets.

The empirical analyses of biofuel developments in the Netherlands, Sweden and Finland confirm the earlier well-established arguments of SNM/MLP and TIS scholars. In addition, we have found three areas, where more research and elaborations could be made in relation to these approaches.

1. *Double carbon lock-in*: MLP explicitly studies socio-technical system transitions and, as underlined in Section 3, it has therefore focused on the destabilization processes of old regimes as a major prerequisite for the success of new technologies. The path dependency literature (see, e.g. Martin & Simmie, 2008; Lovio *et al*., 2011a) has studied the same issue and the notion of carbon lock-in (Unruh, 2000) is widely used in this context. The empirical biofuel analyses reviewed revealed an additional explanation for the carbon lock-in. Not only the old energy regime hampers the emergence of new energy technologies but also the internal difficulties in achieving socio-political acceptance and legitimacy hinder the development and diffusion of many new technologies (biofuels, wind power and nuclear power). Thus, there are two battlefronts for many new energy technologies—a mainly techno-economic efficiency battlefront related to old technologies and a mainly socio-environmental battlefront related to environmental movements and citizen activists. This situation could be called a double carbon lock-in. The recent theoretical emphasis on socio-cognitive and socio-political aspects in the evolution of new energy technologies is useful when examining both of these battlefronts.

2. *Economic dimension*: In Section 4, we stated that there seems to be an underdeveloped area in both approaches. Especially SNM/MLP focuses on socio-technical processes. Yet the role of economic factors, such as oil prices, tax exemptions, feed-in-tariffs and niche market formation was prominent when the evolution of new technologies was empirically explained. The TIS approach includes some economic processes as system functions and Ulmanen *et al.* (2009) suggest that more emphasis should be placed on how financial forms of protection are accomplished by niche actors. In addition, Geels and Turnheim (2010) have divided the external environment of industry actors into two environments, namely task and institutional environments. The task environment is the techno-economic environment where incumbents and new entrants compete on economic grounds. These developments in both approaches signal that the role of economic processes and data may have a stronger role in the future.
3. *Spatial dimension*: In Section 5, the comparison of biofuel development in the Netherlands, Sweden and Finland showed that there are remarkable differences even between these quite similar small EU countries. Differences were explained by endogenous micro-processes and exogenous national characteristics. An additional factor highlighting the importance of the spatial dimension is the variance in dynamics between the EU, the national and the regional/local levels in different country cases. Moreover, the analysis by Sengers *et al.* (2010) found that the axis of global/local benefits/drawbacks is important in explaining media and practitioners' discourses on new energy technologies. Thus, it makes sense to assume that the spatial dimension should be somehow incorporated into analyses at least at the empirical level and that the use of regional and national innovation system literature and geography of innovation literature more broadly could be a relevant source of influence for transition studies in the future.

The TIS and SNM/MLP approaches are well established but still evolving perspectives that have been applied in many empirical studies. However, according to our review the space devoted to data and methods have been surprisingly scant, at least if measured in terms of the length of explicit statements in journal articles. More specific attention to empirical analyses and different types of research designs could provide new insights into how to develop or apply these approaches and also give a good impetus for theoretical elaborations.

In general, our findings indicate that in the area of path creation theories and analyses, at least regarding clean energy technology, a positive atmosphere is evolving. Researchers take new empirical findings seriously and are ready to modify and widen their perspectives based on their observations. Many researchers are looking for crossovers between various path creation approaches. Fruitful discussions between disciplines, such as sociology, science, technology and business studies, geography and economics, are possible and necessary (see also van den Bergh *et al.*, 2011). Yet we do not necessarily argue in the favour of exploring overly complex multi-dimensional frameworks. Rather, we suggest making new efforts to focus on less studied dimensions of path creation theories and complementing findings with different approaches and disciplines.

References

Bergek, A. (2010) Levelling the playing field? The influence of national wind power planning instruments on conflicts of interests in a Swedish county, *Energy Policy*, 38(5), pp. 2357–2369.

Bergek, A., Jacobsson, S., Carlsson, B., Lindmark, S. & Rickne, A. (2008a) Analyzing the functional dynamics of technological innovation systems: A scheme of analysis, *Research Policy*, 37(3), pp. 407–429.
Bergek, A., Jacobsson, S. & Sandén, B. (2008b) "Legitimation" and "Development of positive externalities": Two key processes in the formation phase of technological innovation systems, *Technology Analysis & Strategic Management*, 20(5), pp. 575–592.
van den Bergh, J. C. J. M., Truffer, B. & Kallis, G. (2011) Environmental innovation and societal transitions: Introduction and overview, *Environmental Innovation and Societal Transitions*, 1(1), pp. 1–23.
van Bree, B., Verbong, G. P. J. & Kramer, G. J. (2010) A multi-level perspective on the introduction of hydrogen and battery-electric vehicles, *Technological Forecasting & Social Change*, 77(4), pp. 529–540.
Carlsson, B. & Stankiewicz, R. (1991) On the nature, function, and composition of technological systems, *Journal of Evolutionary Economics*, 1(2), pp. 93–118.
Coenen, L., Raven, R. & Verbong, G. (2010) Local niche experimentation in energy transitions: A theoretical and empirical exploration of proximity advantages and disadvantages, *Technology in Society*, 32(4), pp. 295–302.
Garud, R. & Karnøe, P. (Eds) (2001) *Path Dependence and Path Creation* (Mahwah, New Jersey: Lawrence Earlbaum & Associates).
Garud, R., Kumaraswamy, A. & Karnøe, P. (2009) Path dependence or path creation? *Journal of Management Studies*, 47(4), pp. 760–774.
Geels, F. W. (2002) Technological transitions as evolutionary reconfiguration processes: A multi-level perspective and a case-study, *Research Policy*, 31(8/9), pp. 1257–1274.
Geels, F. W. (2005) The dynamics of transition in socio-technical systems: A multi-level analysis of the transition pathway from horse-drawn carriages to automobiles (1860–1930), *Technology Analysis & Strategic Management*, 17(4), pp. 445–476.
Geels, F. W. (2007) Analysing the breakthrough of rock "n" roll (1930–1970): Multi-regime interaction and reconfiguration in the multi-level perspective, *Technological Forecasting and Social Change*, 74(8), pp. 1411–1431.
Geels, F. W. (2009) Foundational ontologies and multi-paradigm analysis, applied to the socio-technical transition from mixed farming to intensive pig husbandry (1930–1980), *Technology Analysis & Strategic Management*, 21(7), pp. 805–832.
Geels, F. W. (2010a) Ontologies, socio-technical transitions (to sustainability), and the multi-level perspective, *Research Policy*, 39(4), pp. 495–510.
Geels, F. W. (2010b) Towards a socio-evolutionary theory of industrial change: Technology, markets, civil society and polity, *Research Policy*, submitted for publication.
Geels, F. W. & Schot, J. (2007) Typology of sociotechnical transition pathways, *Research Policy*, 36(3), pp. 399–417.
Geels, F. W. & Turnheim, B. (2010) The destabilization of existing regimes in socio-technical transitions: A multi-dimensional framework and case study of the British coal industry (1913–1970). Paper for the Sussex Energy Group Conference, 25th–26th February, University of Sussex.
Geels, F. W., Hekkert, M. P. & Jacobsson, S. (2008) Editorial: The dynamics of sustainable innovation journeys, *Technology Analysis & Strategic Management*, 20(5), pp. 521–536.
Hekkert, M. P. & Negro, S. O. (2009) Functions of innovation systems as a framework to understand sustainable technological change: Empirical evidence for earlier claims, *Technological Forecasting & Social Change*, 76, pp. 584–594.
Hekkert, M. P., Suurs, R. A. A., Negro, S. O., Kuhlmann, S. & Smits, R. E. H. M. (2007) Functions of innovation systems: A new approach for analyzing technological change, *Technological Forecasting & Social Change*, 74, pp. 413–432.
Hillman, K. M., Suurs, R. A. A., Hekkert, M. P. & Sandén, B. A. (2008) Cumulative causation in biofuels development: A critical comparison of the Netherlands and Sweden, *Technology Analysis & Strategic Management*, 20(5), pp. 593–612.
Jacobsson, S. (2008) The emergence and troubled growth of a "biopower" innovation system in Sweden, *Energy Policy*, 36, pp. 1491–1508.
Jacobsson, S. & Bergek, A. (2004) Transforming the energy sector: The evolution of technological systems in renewable energy technology, *Industrial and Corporate Change*, 13(5), pp. 815–849.
Jacobsson, S. & Johnson, A. (2000) The diffusion of renewable energy technology: An analytical framework and key issues for research, *Energy Policy*, 28(9), pp. 625–640.

Jacobsson, S. & Lauber, V. (2006) The politics and policy of energy system transformation—explaining the German diffusion of renewable energy technologies, *Energy Policy*, 34(3), pp. 256–276.
Kemp, R., Schot, J. & Hoogma, R. (1998) Regime shifts to sustainability trough processes of niche formation: The approach of strategic niche management, *Technology Analysis & Strategic Management*, 10(2), pp. 175–196.
Kivimaa, P. (2008) *The Innovation Effects of Environmental Policies. Linking Policies, Companies and Innovations in the Nordic Pulp and Paper Industry* (Helsinki, Finland: Helsinki School of Economics A-329).
Kivimaa, P. & Kautto, P. (2010) Making or breaking environmental innovation? Technological change and innovation markets in the pulp and paper industry, *Management Research Review*, 33(4), pp. 289–305.
Kivimaa, P. & Mickwitz, P. (2011) Public policy as a part of transforming energy systems: Framing bioenergy in Finnish energy policy, *Journal of Cleaner Production*, 19(6), pp. 1812–1821.
van der Laak, W. W. M., Raven, R. P. J. M. & Verbong, G. P. J. (2007) Strategic niche management for biofuels: Analysing past experiments for developing new biofuel policies, *Energy Policy*, 35(6), pp. 3213–3225.
Lovio, R., Mickwitz, P. & Heiskanen, E. (2011a) Path dependence, path creation and creative destruction in the evolution of energy systems, in: R. Wüstenhagen & R. Wuebker (Eds) *Handbook of Energy Entrepreneurship*, pp. 274–301 (Cheltenham, UK: Edwar Elgar).
Lovio, R., Rinkinen, J. & Temmes, A. (2011b) Can biofuels and electric vehicles contribute to the green growth in Finland, in: C. Palmberg *et al.* (Eds) *Towards Green Growth? The Position of Finland in Environmental Technology*, pp. 37–59, Tekes Review 282/2011 (Helsinki, Finland: Tekes – The Finnish Funding Agency for Technology and Innovation).
Markard, J. & Truffer, B. (2008) Technological innovation systems and the multi-level perspective: Towards an integrated framework, *Research Policy*, 37(4), pp. 596–615.
Martin, R. & Simmie, J. (2008) Path dependence and local innovation systems in city-regions, *Innovation: Management, Policy & Practice*, 10(2), pp. 183–196.
Meadowcroft, M. (2009) What about the politics? Sustainable development, transition management, and long term energy transitions, *Policy Sciences*, 42(4), pp. 323–340.
Poole, M. S., van de Ven, A. H., Dooley, K. & Holmes, M. E. (2000) *Organizational Change and Innovation Processes, Theories and Methods for Research* (New York: Oxford University Press).
Raven, R. P. M. J. (2007) Co-evolution of waste and electricity regimes: Multi-regime dynamics in the Netherlands (1969–2003), *Energy Policy*, 35(4), pp. 2197–2208.
Raven, R. P. J. M. & Geels, F. W. (2010) Socio-cognitive evolution in niche development: Comparative analysis of biogas development in Denmark and the Netherlands (1973–2004), *Technovation*, 30(2), pp. 87–99.
Schot, J. & Geels, F. W. (2008) Strategic niche management and sustainable innovation journeys: Theory, findings, research agenda, and policy, *Technology Analysis & Strategic Management*, 20(5), pp. 537–554.
Sengers, F., Raven, R. P. J. M. & van Venrooij, A. H. T. M. (2010) From riches to rags: Biofuels, media discourses and resistance to sustainable energy technologies, *Energy Policy*, 38(9), pp. 5013–5027.
Simmie, J. (2012) Path dependence and new technological path creation in the Danish wind power industry, *European Planning Studies*, 20(5), pp. 753–772.
Suurs, R. A. A. (2009) Motors of sustainable innovation: Towards a theory on the dynamics of technological innovation system, PhD Thesis, Innovation Studies Group Copernicus Institute, Utrech University, Utrecht, the Netherlands.
Suurs, R. A. A. & Hekkert, M. P. (2009) Cumulative causation in the formation of a technological innovation system: The case of biofuels in the Netherlands, *Technological Forecasting & Social Change*, 76(8), pp. 1003–1020.
Suurs, R. A. A., Hekkert, M. P., Kieboom, S. & Smits, R. E. H. M. (2010) Understanding the formative stage of technological innovation system development: The case of natural gas as an automotive fuel, *Energy Policy*, 38(1), pp. 419–431.
Ulmanen, J. H., Verbong, P. J. & Raven, R. P. J. M. (2009) Biofuel developments in Sweden and the Netherlands—protection and socio-technical change in a long-term perspective, *Renewable and Sustainable Energy Reviews*, 13(8), pp. 1406–1417.
Unruh, G. (2000) Understanding carbon lock-in, *Energy Policy*, 28(12), pp. 817–830.
Van de Ven, A. H., Polley, D. E., Garud, R. & Venkataraman, S. (1999) *The Innovation Journey* (New York: Oxford University Press).
Wüstenhagen, R. & Wuebker, R. (Eds) (2011) *Handbook of Energy Entrepreneurship* (Cheltenham, UK: Edward Elgar).

Renewable Energy Technology and Path Creation: A Multi-scalar Approach to Energy Transition in the UK

JÜRGEN ESSLETZBICHLER

Department of Geography, University College London, Gower Street, London WC1E 6BT, UK

ABSTRACT *This paper examines the potential contribution of UK regions for developing and deploying renewable energy technologies to achieve the government target of obtaining 20% of its energy from renewable sources by 2020. The paper argues for a multi-scalar approach to energy transition theory and policy. National-scale processes and policies need to be complemented by regional and local policies in order to discover and incorporate meso-level sources of renewable energy, recognize that niche or path creation is a geographically localized process and mobilize heterogeneous, local actors around common "regional energy visions" to improve implementation of renewable energy projects. After critically reviewing the main theoretical approach to energy transitions, the multi-level perspective, the paper employs patent data to describe the comparative position of UK regions in the renewable energy sector and examines the success of Danish, German and Spanish regions resulting from strong government intervention at the national level supplemented by region-specific strategies. A number of policy strengths and shortcomings are identified in the evolutionary trajectory of the UK energy system including weak technology push and policy pull factors. Finally, the paper reviews existing regional renewable energy policy and speculates on the potential impact of recent changes in spatial and energy policies on the ability to deploy and develop renewable energy sources in the UK.*

Introduction

In order to limit average global temperatures to 2°C above pre-industrial levels, global carbon emissions need to be limited to a cumulative total of 750 Gt CO_2 between 2010 and 2050 (Broeckner, 2007; Messner *et al.*, 2010). The key driver to approximate this goal is the decarbonization of the energy sector. Assuming a modest growth rate of 1.6% per year between 2008 and 2050 implies an electricity supply of 39,000 TWh in 2050 (Jacobsson & Bergek, 2011). Given the currently installed capacity of renewable and nuclear energy sources of 6600 TWh (International Energy Agency, 2011) entails the installation of additional renewable capacity of 32,000 TWh which is five times the capacity of the whole power sector in 1973. The task is enormous and requires the search for new technologies and implementation of all existing renewable energy technologies (RETs) immediately with the goal of a transition from carbon to RETs in a relatively short period of time. The UK has committed to supply 20% of its energy from renewable sources by 2020 demanding rapid transition from a fossil-fuel-based energy system to one powered by renewable sources.

The dominant theoretical framework on transition (Geels, 2002; Smith *et al.*, 2010) is strongly influenced by evolutionary economics, in particular, the work by Neo-Schumpeterians on technology regimes, paradigms and trajectories (Dosi, 1982; Freeman & Perez, 1988), the systems of innovation literature (Lundvall, 1992; Nelson, 1993) and work on co-evolution (Norgaard, 1994; Nelson, 2001), and proposes a multi-level perspective (MLP) to conceptualize transitions where new technological pathways are created in socio-technical *niches* operating outside the dominant socio-technical *regime* that are also influenced by *landscape* processes such as resource scarcity, peak oil, energy security or economic crisis. The theoretical framework has subsequently been broadened to recognize the interdependence between the evolution of the technological, economic, environmental, social and political subsystems with strong recent emphasis on actor-network and discourse analysis of existing policies (Shove & Walker, 2007; Walker, 2008; Walker & Devine-Wright, 2008). This paper will critically review this literature and identify the strengths and a number of shortcomings with the approach, in particular, the lack of attention to spatial scale, and employ insights from evolutionary economic geography to argue for a multi-scalar analysis of transitions in general, and energy transitions in particular.

MLP is employed to examine the transition towards renewable energy supply in the UK with a particular emphasis on the regional scale that has received little attention in the literature (for an exception, see Smith, 2007). As regional transition processes are linked to decisions taken at the national scale, it is impossible to understand the UK's energy transition without first discussing niche development and landscape pressures for the UK as whole. For this purpose, the UK is compared with the most successful countries in the European Union (EU): Denmark (DK), Germany (GE) and Spain (SP). While causality is difficult to establish, DK, GE and SP are all characterized by successful regional clusters in RETs that are, to some extent, driving and accelerating national transition processes. In order to understand the lack of strong regional participation and leadership in the transition process in the UK, it is necessary to examine regional renewable policies.

The paper addresses those broad concerns by answering the following research questions: First, why should MLP conceptualize transitions on multiple spatial scales? Second, can differences in niche development and national policy landscapes account for the relatively low renewable energy share in the UK's energy portfolio? Third, how can regional transition processes and policies complement and strengthen national transition processes? Fourth, how does the innovative potential of UK regions compare with those of other European regions? Fifth, what regional renewable energy policies are in place to stimulate deployment of and niche creation in renewable energy policies in the UK?

In order to answer these questions, section two critically reviews the MLP and highlights the theoretical importance of a multi-scalar approach to energy transitions in order to increase deployment levels and exploit local externalities for niche creation in renewable energy sources. Section three provides the empirical analysis. Utilizing secondary data and various UK government documents, the paper will first compare deployment shares, patent activity, R&D spending and incentive structures for renewable energy sources in the UK with the leading EU countries such as DK, GE and SP at the national scale revealing the importance of strong market incentives and technology push for renewable energy deployment. Second, in order to identify the capacity for regional path creation, the number of patents in RETs in UK regions is compared with those of other European regions. While national policies are clearly important, there are a number of cases where regional policy intervention resulted in strong regional clusters of renewable development and deployment that may, in some cases, accelerate development and deployment at the national scale. Third, the case of Navarra will be used to illustrate that strong economic incentives and political leadership can accelerate transition processes in the region but also influence the process at the national scale. Regional champions in RETs that could serve as showcase and influence national transition processes are currently missing in the UK. Fourth, the weaknesses of past regional policies to stimulate deployment and path creation in RETs are summarized and the potential merit of recent changes in energy, economic and spatial policies for increasing regional deployment, innovation and production of RETs in the UK are examined. Section four concludes the paper.

Conceptualizing Energy Transitions: The MLP

There are at least four approaches to research transition processes: innovation systems, MLP, complex systems and evolutionary systems (Van den Bergh *et al.*, 2011). While they emphasize different theoretical concepts and prioritize different empirical characteristics of transitions, they all focus on the relationship between the creation of diversity in form of innovation and niches and selection of diversity through context-specific market, institutional, social and cultural pressures. Probably the most widely used concept to study socio-technological transitions is the MLP. The MLP builds on the systems of innovation approach, evolutionary economics and sociology of technology. The systems of innovation approach is useful for characterizing and identifying the key components and (primarily) technical relationships among existing system components, but has little to say how the system can be transformed into something else (Edquist, 1997; Coenen & Diaz Lopez, 2010). Enriching the systems of innovation approach with insights from evolutionary economics and the sociology of technology adds a dynamic component and broadens the systems perspective to include social, institutional and multi-actor networks in the analysis.

The MLP organizes analysis of transitions into a socio-technical system that consists of *niches* (micro-level), *regimes* (meso-level) and *landscapes* (macro-level) (Kemp, 1994; Rip & Kemp, 1998; Geels, 2002, 2005; Geels & Schot, 2007; Smith *et al.*, 2010). The starting point for the development of the MLP was Nelson and Winter's (1977) notion of a technological regime referring to the cognitive rules and design heuristics that guide individual and firm searches for and development of innovations along specific technological trajectories. Rip and Kemp (1998) and Geels (2002) broadened the concept insisting that technological regimes are embedded in institutions and infrastructures that shape technological trajectories. Geels (2002) replaced the term technological with *socio-technical regime*. More specifically, Geels (2005) suggests that the regime consists of three interlinked dimensions: "(1) networks of actors and social groups; (2) formal, normative and cognitive rules that guide the activities of actors; (3) material and technical elements" (Verbong & Geels, 2007, p. 1026). In the UK electricity regime important actors are utilities, large industrial users, households and the Department of Energy and Climate Change (DECC) (Foxon *et al.*, 2010). Formal rules include regulations, standards and laws. Cognitive rules are belief systems, problem agendas, guiding principles and search heuristics. Normative rules encompass the role of relationships and behavioural norms. Material and technical elements includes resources, grid, power generation plants and so on.

Existing social networks represent "organizational capital" that is protected and mobilized by powerful, incumbent actors to serve their interests. Regulations and standards may stabilize regimes and strong ties among actors may result in cognitive lock-in that prevent actors from searching for solutions outside the regime (e.g. for renewable energy sources, decentralized energy production, etc.). Investments are sunk in large-scale infrastructure that is protected from devaluation by its owners and network complementarities result in increasing returns, further strengthening the status quo. For these reasons, the dominant regime in general and the energy sector in particular exhibits strong path dependence and lock-in (Hughes, 1983; Grabher, 1993; Arthur, 1994; Unruh, 2000; Scrase *et al.*, 2009).

As a result of lock-in, socio-technical regimes tend to produce complementary, incremental innovations that strengthen existing technological trajectories. Radical innovations that may form the technological basis for new path-breaking, socio-technical regimes are generated in *technological niches*. Radical innovations are thought to need protection because they are often expensive, less efficient and not complementary with existing socio-technical regimes. Universities, the armed forces, social networks of entrepreneurs, alternative think tanks or region-specific selection environments may provide a protective umbrella from immediate market selection. As niche innovations provide the seeds for transition, niche development needs to be nurtured, stimulated and managed. Niche expansion is dependent on a supportive institutional framework and socio-cultural environment, co-evolution of complementary technologies and the establishment of strong networks among niche and component producers and users to generate the necessary externalities to achieve economic viability and eventually superiority over regime technologies (Hekkert & Negro, 2009). Assuming that socio-technical regimes are complex adaptive systems, it is important that a portfolio of niches is maintained to obtain both, design and path flexibility (Alhemade *et al.*, 2009; Stirling, 2011; Van den Bergh *et al.*, 2011).

Socio-technical regimes are situated in a *socio-technical landscape* or selection environment that contains a number of heterogeneous factors such as oil prices, economic growth, wars, demographic trends, political coalitions, cultural and normative values, or environmental problems but also government regulations. Landscapes serve as external, slow-changing environments that exert pressure on regimes and niche production. Rates of economic growth influence investment patterns that in turn, influence inventive activity and/or diffusion of alternative technologies. Changes in the institutional environment influence the search for and selection of alternative sources of energy production and distribution. Regional and national governments can shift selection pressures in favour of RETs through regulatory and financial instruments such as renewable obligations (ROs) or feed-in-tariffs (FITs), public R&D funding or carbon taxes. Growing environmental awareness as part of socio-cultural development may be considered a landscape process that puts pressure on politicians, firms, organizations and opens up windows of opportunities for niche alternatives to compete with dominant technologies.

The MLP offers a relatively straightforward way to organize and simplify complex transition processes. Its conceptual framework links specific innovation activities in niches to large scale transformations at the regime and landscape levels and so mirrors the relationship between variety/path creation and selection of evolutionary theory. The detailed historical analysis of niche, regime and landscape change to explain transitions by the MLP offers a rich picture of how new socio-technical regimes rather than a single technology or sector (which is often the focus of evolutionary economics) emerge and eventually become dominant.

> The MLP argues [that] portfolios of policy measures need to work across: the destabilization [unlocking] of incumbent regimes (so that opportunities for structural change increase); the promotion of radical green niches (so that the portfolio of promising solutions broadens); and processes for translating ideas and practices from niches into mainstream settings. (Smith *et al.*, 2010, p. 445)

Despite its analytical appeal and relative success in empirical research, a number of criticisms have been launched against the MLP (Smith *et al.*, 2010). First, the relationship between niche, regime and landscape levels are far more complex than originally envisioned by the the MLP. Second, the interaction between plural regimes and niches are more complex than suggested by the MLP. Third,

> making the core concepts of niche, regime and landscape operational for empirical research is a question of bounding, partitioning and ordering the system under study. Any attempt at bounding and analyzing complex, emergent socio-technical systems will necessarily be partial, situated and temporary. (Smith *et al.*, 2010, p. 444)

Fourth, as transition researchers are attempting to actively manage the transition process, a clear understanding of existing power relations and distribution of resources among various actors is necessary. A fifth shortcoming and the most important in the context of this paper is the lack of attention to spatial scale. The MLP generally operates at the national scale because most of the regulations, institutions and policy audiences operate at that scale, but regional niche creation and landscapes may complement, speed up or even shape national transition processes.

Regionalization to Unlock National Energy Regimes

Because the majority of existing energy regimes are centralized, decentralization of energy production will result in frictions in the existing regime necessary for regime transitions to occur. Decentralization of the energy system would increase flexibility, result in locally specific energy systems challenging existing power relations and oligopolies, devalue existing infrastructure arrangements and social capital and result in a modular arrangement of energy supply and use. Devaluing infrastructures and social capital could lower profit rates of national suppliers, grid operators and generators, increase energy prices (and costs for consumers) and level the playing field opening up windows of opportunities for niche developments and force incumbents to adapt to changing circumstances.

An exclusive focus on national energy generation at the expense of regional energy regime development also under-utilizes a number of potential sources for renewable, low-carbon energy solutions such as district heating (Watson *et al.*, 2010). Learning effects from the *deployment* of region appropriate renewable energy sources are likely to stimulate the *development* of renewable energy (e.g. Danish energy cluster Fiona supporting among others, the formation of a district heating centre of knowledge[1]). Exploitation of community energy sources is vital to transform energy sectors. In this sense there is no option but to decentralize the grid (Unruh, 2000).

How regions plug into and transform the existing energy system will depend on regional characteristics. In the UK, the recently commissioned regional renewable capacity studies identify the potential of various renewable energy sources at the regional and local authority levels (DECC, 2011[2]). Locally and regionally specific solutions are necessary because locally existing infrastructure configurations such as housing, transportation, energy systems and skill sets as well as "natural" environments for energy generation vary and need to be tailored to reflect those differences. Examples include "hydrogen cities" mobilizing their relational assets and focusing on technological fixes to reduce energy use and carbon emissions (Hodson *et al.*, 2008; Hodson & Marvin, 2010) as well as rural "energy regions" developing energy visions around local resource use (e.g. biomass), energy autarky and regional energy surplus creation (Späth & Rohracher, 2010). National energy regimes can be challenged through this increasing diversity of smaller energy systems based on the deployment of different renewable energy sources, some of which will necessitate the development of smart grids.

While deployment of renewable energy systems serves regions directly through the ability to reduce energy import and generate revenue from sales of energy, regional leakages of surplus value can be reduced if actors learn from installing and using renewable energy systems and contribute themselves to niche creation.

Regional Path Creation

Unlocking the national energy system opens up the potential for developing RETs often available in green niches. If, as evolutionary economic geographers claim, knowledge spillovers are geographically constrained (Jaffe *et al.*, 1993) and innovation is a geographically localized process stimulated by spin-off dynamics, network formation and labour mobility resulting in the creation of virtuous innovative cycles (Boschma, 2009), then the regional dimension has to be developed explicitly in any successful transition to a low-carbon economy where niche innovation in "green technology" forms an essential part (HM Government, 2009).

Evolutionary economic geography is successful in explaining path-dependent development and lock-in once new niches or paths have been created, but attributes the actual path creation process to "historical accidents". This is true for the windows of locational opportunity approach (Boschma & Van der Knaap, 1997), models of path dependence (Arthur, 1994) and spin-off dynamics (Klepper, 2002). Martin (2010) proposes an alternative model of path dependence with particular emphasis on the pre-formation and path creation phase where the existing local economic and technological structures, knowledges and competences are mobilized through the purposive actions of agents resulting in the local emergence of new paths. Endogenous creation by local actors, the adoption and adaptation of novel extra-regional knowledge through FDI or firm-internal knowledge pipelines and the recombination of existing knowledge bases may result in new path creation (Martin & Sunley, 2006; Neffke *et al.*, 2011). Changes may also be triggered by strong selection (landscape) pressures (such as new environmental regulations) that reshuffle national and local efficiency hierarchies (Essletzbichler & Rigby, 2007).

The application of niche innovations is often outside the core, suggesting that the opening of a technological window of opportunity may be coupled with local path creation. Niche development could thus be combined with regional development in non-core regions (Scott, 1988; Storper & Walker, 1989; Boschma & van der Knaap, 1997). Regional path creation based on clean energy technology will not only enhance national competitiveness of the industry but is also likely to benefit the regions. Clean energy technology has long-term market potential because it requires the complete reconfiguration of existing energy systems and is likely to deliver jobs, skills and revenues that benefit the local economy (Innovas, 2009).

A successful transition requires the enrolment of heterogeneous actors to sign up to a common vision in order to secure cooperation, strengthen cluster development and increase the effectiveness of transition policies. These visions will differ by place and are thus easier to develop if they build on needs of actors and availability of resources in those places.

Recent work on transition research highlights the importance of "soft" social and cultural factors to trigger and develop transition pathways (Smith *et al.*, 2005). Späth and Rohracher (2010) demonstrate how sustainable Austrian "energy regions" can serve as "discursive niches" of transitions to sustainable low-carbon energy systems. In the case of the "energy region Murau", a rural area consisting of about ten municipalities, the key objectives were energy autonomy, export of energy carriers resulting in economic surplus and a high level of public awareness concerning the need for an energetic circular flow economy. Linking environmental sustainability to economic development was important in order to enrol strategic actors, such as businesses, in the alliance. Emphasis was put on potential for value creation by bringing regional resources into use with biomass being the obvious choice. Workshops and subgroups of actors developed concrete objectives focusing on issues such as wood-fired district heating systems, solar heating systems for private homes, projects for renewable electricity production and ways to improve the energy efficiency of buildings. Frequent meetings facilitated the realization of various concrete projects and significantly reduced transaction costs. Changes in planning laws, development of specific technologies (e.g. new truck design to deliver wood chips), coordinated R&D efforts, logistical infrastructure (biomass bourse) and services resulted in improvement of technical solutions and business strategies that could be exported to other regions. There was significant support from the meso-level for this project but the success story was then also employed as "proof of principle" to initiate transformations in other places and at a higher spatial scales. The example demonstrated that "energy solutions" can operate at any scale and any type of region and that successful solutions can be exchanged, adapted, diffused and probably scaled up to actively transform the national energy regime.

Landscape Pressures

A spatial focus on energy production and consumption could alter the selection pressures if regions are allowed to manipulate their selection environments (i.e. R&D policies, sustainable energy and low-carbon mandates, different tax and incentive structures). Regional institutional umbrellas could preserve inter-regional variety in niche development and thus maintain higher levels of diversity at the niche level. A spatial focus would also take into consideration the locally or regionally existing resource and infrastructure, firm, knowledge and skills bases (Martin, 2010).

Landscapes exert various pressures on regimes, regime components and niches and could be conceptualized as multiple selection environments operating on various spatial scales (Essletzbichler, 2012). Regions are subject to national landscape pressures while regional regimes are also subject to regional selection pressures. There are different ways for national governments to alter the selection environment and exert pressure on regions. Taking into account the results from capacity studies, emission budgets could be devolved to regions or local authority level. Regions successful in generating a large share of renewable energy would then be able to trade carbon permits with regions that are unable to do so. The policies may lead to transfer payments from cities to rural areas, the attraction of "green" industries and services with low-carbon footprints or the "greening" of existing industries. Revenues can then be reinvested to strengthen regional niche developments (as in the case of Southampton's geothermal energy generation project).

Regional landscapes can be designed to redistribute resources from high-carbon to low-carbon industries and firms within the region, that is, regions would have the ability to alter the price structure facing businesses and industries to push them in directions that mesh with regional environmental objectives. Regional selection environments can be designed around normative goals to influence consumer behaviour, which in turn, will influence firm behaviour. Tradeoffs among various energy saving and renewable energy generation projects can then be negotiated influencing the regional transition pathways.

If we accept the premise that attention to scale is important for the development of transition policies because regions exploit niche markets (such as medium-sized wind farms, district heating systems, energy conservation programmes, land-use planning changes) that are excluded by the incumbent regime, diversity in niche production can result in cross-regional learning as well as upward and downward scaling of existing regime structures to enhance renewable capacity generation and deployment. Hence, the transition model needs to take spatial scale into account explicitly. A visual representation of a multi-scalar, multi-level transition approach is depicted in Figure 1.

The key message here is not that regions function as independent regimes or that the regional scale is more important than the national scale but that regions, if properly endowed with regulatory authority and monetary means, contribute to the creation and maintenance of diversity in the niche practices and discourses that influence regional regimes that may or may not be scaled up, replace, interact, transform or be taken over by national regimes or other "fitter" regional regimes. Because regions differ in terms of their infrastructure, natural resource endowments, social and cultural capital, region-specific transition governance is not only desirable but necessary to improve the probability of success to a transition to a low-carbon economy. This reasoning opens up the possibility of multi-scalar transition research where transition processes at the regional scale co-evolve with processes at the national scale.

While the theoretical discussion suggests an important role for regions to develop niches and accelerate transitions at the national scale, the translation of theoretical categories such as niche, regime and landscape into empirical categories is imperfect (Smith *et al.*,

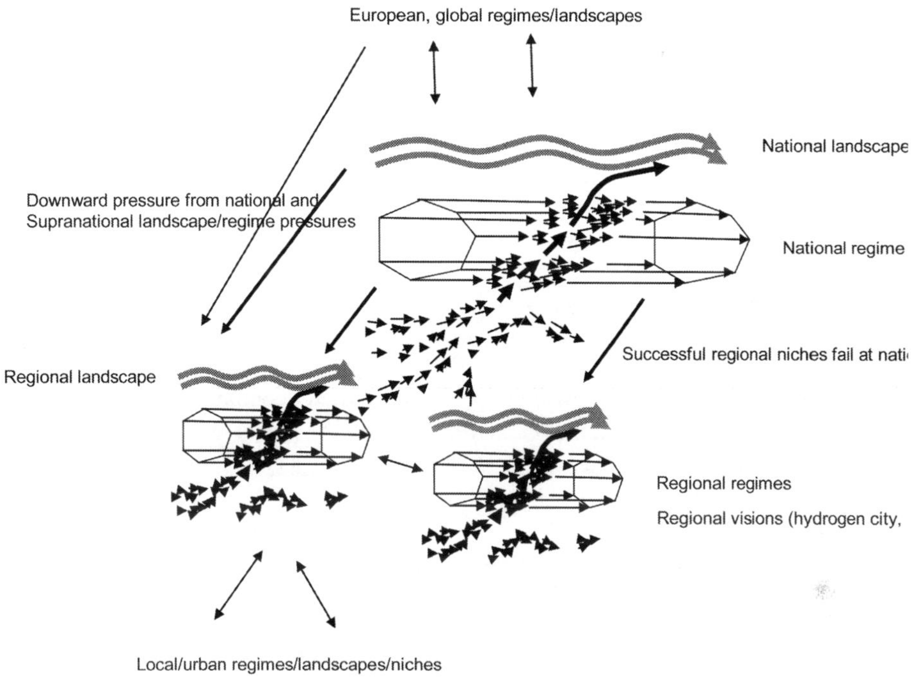

Figure 1. Multi-scalar MLP.
Source: Based on Geels (2002).

2010). In this paper, the potential for niche creation in RETs is approximated with patent data at the regional and national scale while the relative success of moving to a sustainable energy regime can be approximated by the share of renewable energy on gross energy and/or electricity consumption. Landscape pressures expected to impact the deployment and creation of RETs are defined narrowly as R&D policies and spending, investment, price and tax incentives as well as regional and local planning guidelines. In order to gain a better understanding of key factors driving energy transitions, the UK is compared to the relatively successful countries of DK, GE and SP[3].

In order to answer the research questions posed in the introduction and get an overview of the key differences and similarities of innovation and deployment in renewable energy between those countries, the following empirical part will first compare deployment shares, patent activity, R&D spending and incentive structures at the national scale. Second, patents in RETs of EU regions are mapped and relative strengths and weaknesses of UK regions summarized. The case of Navarra is discussed in greater detail because it demonstrates how regions can take the initiative and influence national transition processes. Third, existing regional policies to develop and deploy renewable energy sectors in the UK are summarized and the potential impact of current changes to energy and spatial policies on the transition process discussed briefly.

Empirical Analysis

The empirical analysis is based on secondary data sources, including patent data from the OECD, R&D expenditure data and information on incentive and tax policies as well as legislation regarding RETs from the International Energy Agency (IEA), and information on deployment levels from Eurostat. Unfortunately, official industry and export statistics could not be used as they do not discriminate between renewable energy and fossil-fuel technologies. The UK regional renewable policy is based on various government documents. A brief discussion of the data sources follows next.

Data

Eurostat[4] provides various indicators on the shares of renewable energy. For this paper, the share of renewable energy on total gross energy and total gross electricity consumption was used to get an indication of the success of national energy transition processes (Figure 2).

Patent data are generally interpreted as proxy for inventive output and innovative input (OECD, 2009). At the national scale, OECD[5] provides patent data for a number of environment-related patents while "renewable energy" patents at the TL3[6] regional level are not disaggregated further. Following the advice of OECD (2009), this analysis uses Patent Cooperation Treaty applications at the international phase (to control in part for home bias of applicants), priority year (to count patents close to the date of invention), fractional counts and allocates patents to countries/regions based on the location of the inventor rather than the applicant. This is sensible because the location of the inventor reflects where research is actually taking place rather than where the applicant (often a company) is headquartered. Patent data are not without problems. They include applications rather than actually granted patents, they do not distinguish between the relevance of applications and often contain several, non-hierarchical patent codes. If there are several renewable energy codes on the application, then the patent may be counted more than once.

Invention and innovation will depend in part on technology push and in part on market pull factors. In order to gain an idea of how strong national government support research in RETs, renewable energy R&D data from the IEA are employed. Only public R&D expenditures are included, which may bias results in favour of those countries with nationalized energy regimes. Because energy produced from renewable sources (with the exception of onshore wind) tends to be more costly than conventional energy sources (DECC, 2011), incentives need to be provided to increase deployment and potentially innovation in RETs. Data on different incentive, tax and regulatory systems are also provided by the IEA and Haas *et al.* (2011) and will be used to compare differences in national policy frameworks.

And finally, in order to examine the role of current and future regional polices to support the energy transition in the UK, various UK government documents from the DECC, the Department of Business, Innovation and Skills (BIS) and the Department for Communities and Local Government as well as various regional development agencies (RDAs) are used.

National Scale

To set the scene, Figure 2 depicts the renewable shares in gross energy (Figure 2(a)) and electricity (Figure 2(b)) consumption in the four countries and the EU-27 average. The 2009 renewables' share on gross energy consumption for EU-27, DK, GE, SP and the UK were 9.0%, 16.7%, 8.5%, 9.3% and 3.0%, respectively. Their renewables' shares in electricity con-

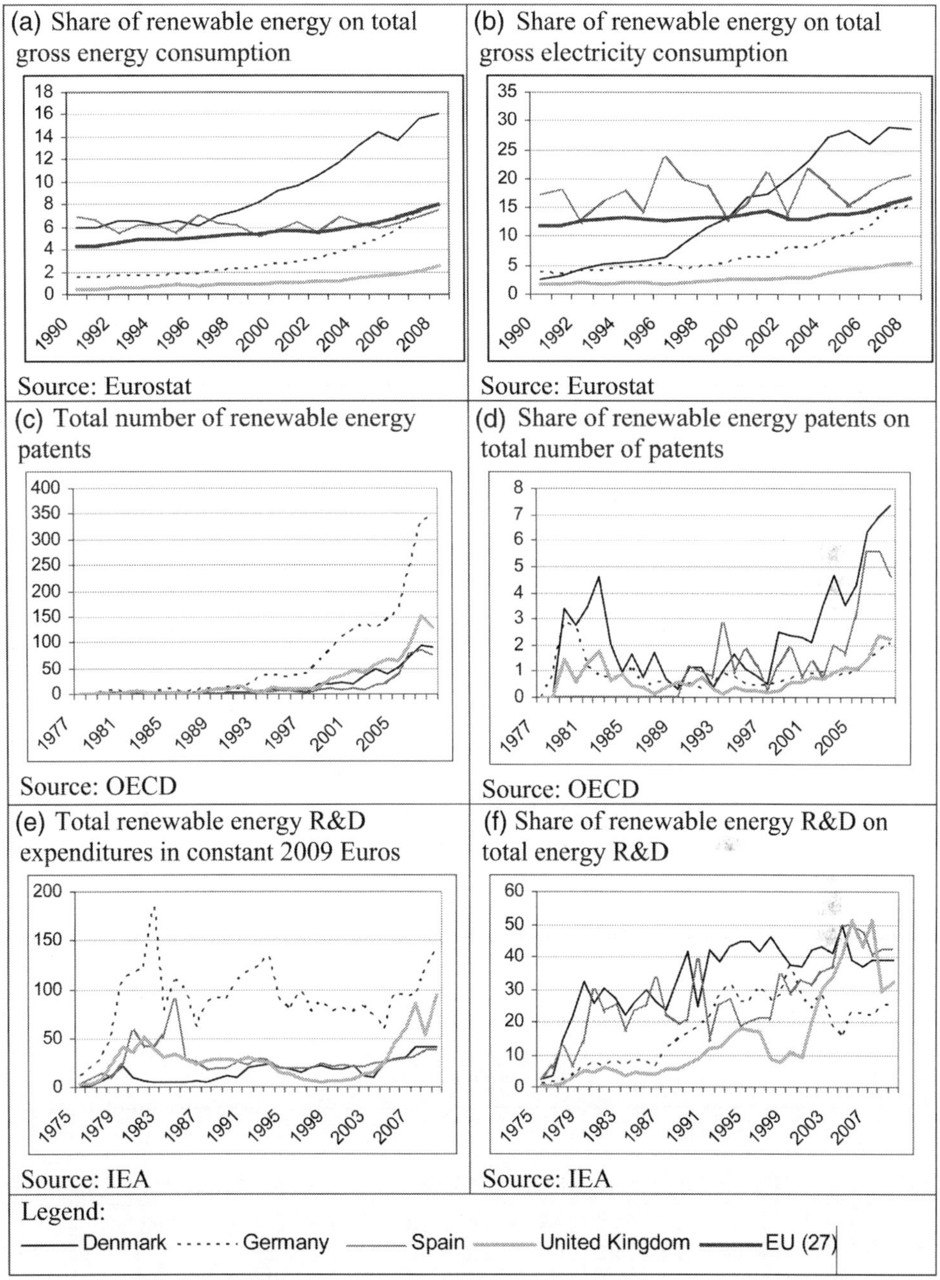

Figure 2. Deployment, patents and R&D trends for DK, GE, SP and UK.
Source: Compiled from Haas *et al.* (2011) and IEA (2011).

sumption in 2008 were 16.7%, 28.7%, 15.4%, 20.6% and 5.6%, respectively. Figure 2 reveals that the UK is far behind the leading EU countries but even behind the EU-27 average in achieving a transition from a fossil-fuel-based to a renewable energy system.

In order to get a better understanding of the relative failure of the transition process in the UK, it is useful to compare innovation in and the regulatory structure guiding renewable energy sources in the UK with the experiences of the relatively successful countries such as DK, GE and SP. According to the MLP, we need to identify green niche development and landscape pressures that may facilitate or impede the creation and diffusion of those niches. Drivers at the national scale are examined first. Figure 2(c) plots the number of patents in all RETs from 1976 to 2008. The number of patents in this sector remained relatively low until the early 1990s. GE has the highest number of patents while DK and SP emerge as leading countries in RETs once we standardize by the total number of patents (Figure 2(d)).

Table 1 presents the cumulative number of patents and the revealed technological advantage (RTA) for each of the renewable energy sectors for the period 2000–2008. RTA measures the share of renewable patents on the total number of patents in a country divided by the share of renewable patents in all countries on the total number of patents in all countries and is often considered as an indicator of comparative innovative advantage of a country in a particular technology sector. An RTA > 1 means that a country patents more in this sector than would be expected based on its total number of patents. DK's strong performance and concentration in the wind power sector is confirmed by an RTA of 14.1. GE holds the largest amount of patents in all renewable energy sectors save for marine and tidal/wave technology although its RTAs tend to be smaller than in other countries because of its large number of total patents. SP is now holding significant numbers of patents in wind, solar thermal, solar photovoltaic (PV) and wave technologies. The UK has a comparative advantage in marine and wave/tidal technologies and also improved its position in the wind power sector (possibly due to its competitive position in offshore wind).

Inventive output is driven in part by R&D expenditures and in part by market expansion. Figure 2(e) traces the evolution of public R&D in renewable energy sectors. After fast increases in R&D budgets as a reaction to high energy prices in the 1970s, R&D expenditures dropped during the recession of the early 1980s. GE had the highest R&D budget, while those of the UK, SP and DK were relatively similar for most of the period despite the UK economy's size (see Jamasb *et al.* (2008) on UK's R&D policy). DK and SP started

Table 1. Patents, shares and RTA for DK, GE, SP and the UK.

	DK			GE			SP			UK		
Energy source	Patents	Share	RTA	Patents	Share	RTA	Patents	Share	RTA	Patents	Share	RTA
Renew. all	401.6		3.9	1486.1		1.0	308.4		3.0	584.3		1.0
Wind	341.4	81.1	14.7	497.1	29.6	1.5	134.8	38.5	5.7	167.1	24.6	1.3
Solar thermal	26.0	6.5	1.7	274.3	16.3	1.3	102.1	29.1	6.8	65.5	9.6	0.8
Solar PV	9.1	2.3	0.2	698.3	41.5	1.1	61.3	17.5	1.3	232.5	34.2	0.9
Solar hybrid	1.0	0.2	0.7	32.3	1.9	1.6	4.0	1.1	2.7	5.7	0.8	0.7
Geothermal	3.0	0.7	1.3	47.0	2.8	1.4	2.0	0.6	0.8	9.5	1.4	0.7
Marine (excl. tidal)	20.5	5.1	4.6	20.9	1.2	0.3	23.0	6.6	5.1	76.9	11.3	3.2
Tidal, stream	5.2	1.3	1.4	31.8	1.9	0.6	7.0	2.0	1.9	72.2	10.6	3.6

Source: Data based on OECD Patent Database, http://www.oecd.org/sti/inno/oecdpatentdatabases.htm

investing substantial shares of their energy R&D budgets in renewables in the late 1970s and early 1980s (about 40% of their total budgets) (Figure 2(d)). DK focused its resources on wind and bio-energy early on, while SP diversified its investment portfolio into geothermal (until the late 1980s), bio- and solar energy with wind playing only a minor role until the late 1990s (Klaasen *et al.*, 2003). GE's renewable R&D shares started to increase in the late 1980s with PV and, since the mid-1990s, wind being the main beneficiaries. The UK was characterized by a low percentage of renewable R&D on total energy R&D until the 2000s despite early investment in ocean and geothermal energy research. The UK's renewable share on total energy R&D was less than 10% until the 1990s and less than 20% until 2000 when it started to climb to 50% in 2006. The recent increases in R&D expenditures are primarily due to an increase of R&D investment in the PV sector but may also reflect commitment to offshore technologies. Econometric analyses show that public R&D expenditures have a positive and statistically significant effect for patenting activity in the wind and solar energy sector (Johnstone *et al.*, 2010) while work on the relative effects of induced technological change and policy-induced substitution suggest that the latter is more important for generating investment in renewable technologies (Hascic *et al.*, 2010).

While R&D and patenting activity in the UK appeared relatively low compared to the other three countries, the renewable deployment gap is even wider. It is thus necessary to look at the evolution of the incentive and regulatory landscape in those countries to understand how diffusion of niche technologies has been accomplished. Figure 3 summarizes the main policy changes affecting research in and deployment of RETs including investment incentives, tax credits, low interest/soft loans, tendering systems for investment

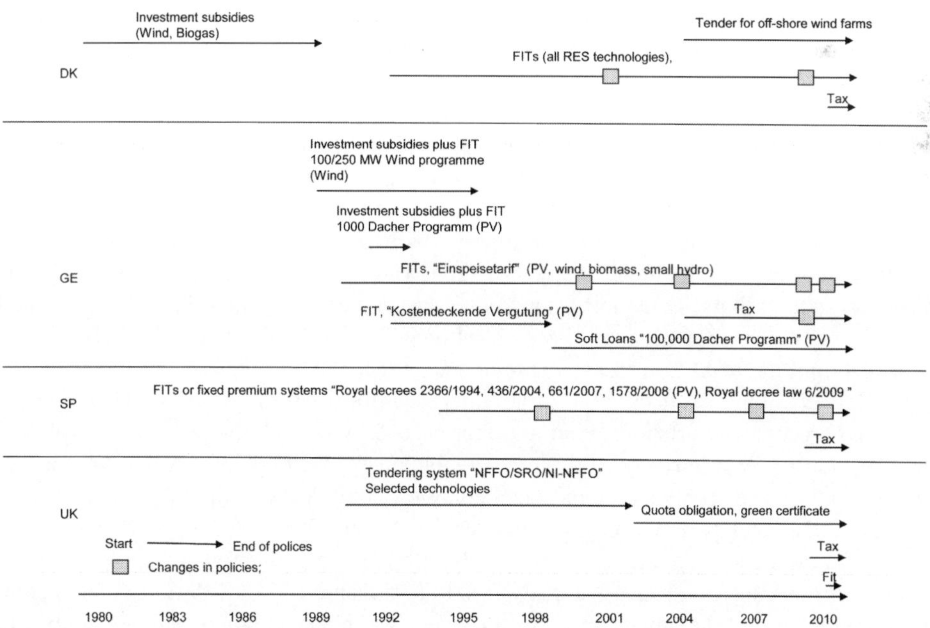

Figure 3. Key policies with respect to renewable energy in Denmark, Germany, Spain and UK.
Sources: Compiled from Haas *et al.* (2011) and IEA (2011).

grants, FITs and fixed premium systems, tendering systems for long-term contracts and tradeable green certificate systems (for a summary of the various instruments, see Haas *et al.*, 2011).

Wind energy was promoted in DK since the 1970s through a mixture of investment subsidies, tax refunds, R&D support and, since the 1990s, FITs (Klaasen *et al.*, 2003). Up to the 1990s, the main owners of wind turbines were cooperatives resulting in high deployment rates and low objections to planning applications (Munksgaard & Morthorst, 2008; Haas *et al.*, 2011). The careful timing between R&D support, investment incentives and procurement support played an important role in the success of the Danish wind industry with Vestas dominating the global market for wind turbines (REN21, 2011) and the Aarhus–Aalborg cluster attracting Siemens, Suzlon and Gamesa to DK (Cooke, 2010). GE has encouraged the use of wind energy since the 1970s through the use of FITs and tax breaks, resulting in a rapid diffusion of wind power (Klaasen *et al.*, 2003), while the rapid increase in solar PV was supported through investment subsidies, low interest loans and FITs. The Spanish renewable energy sector benefited primarily from generous FITs that resulted in rapid deployment of wind and more recently PV. SP has the second largest installed wind capacity and second largest PV capacity and is the largest solar thermal electricity producer in Europe. The regulatory framework provided stability for developers to recover costs necessary for the rapid deployment of those sources (IDAE, 2010). In contrast to the above three countries, the UK relied primarily on tendering, green certificates and ROs. The tendering system and ROs were useful for obtaining low support prices but only the best, most competitive locations were viable while renewable energy targets were never met (Haas *et al.*, 2011). As a result of those policies, the UK lags far behind the European leaders in renewable energy generation and innovation.

This is now recognized and a number of recent changes to promote renewable energy in the UK have been introduced. The most important change affecting the renewable energy sector is the introduction of an FIT system to complement and eventually replace the RO system (DECC, 2011). Since then, 164 MW of capacity was installed, of which 121.6 MW are small-scale solar PV installations (DECC, 2010). Unfortunately, the first review of the FIT took place in February 2011, causing uncertainty and instability in the market. The implementation of FITs is complemented by investment in the grid to accommodate variable power sources such as wind and solar energy but also to connect offshore wind farms to the onshore grid, the green deal to improve building efficiency, the micro-generation strategy and, as overarching strategy, the "renewable energy roadmap" (DECC, 2011) that sets out various strategies to increase deployment of renewable resources and implementation is supported by the six National Policy Statements (NPS) for Energy that highlight the need to increase deployment rates for renewables and provides guidelines for planning decisions on large energy infrastructure projects. The "energy roadmap" recognizes offshore wind and marine energy as technologies to be exploited by UK industry. While rising R&D in renewable energy, strong competitive advantages in offshore (wave, tidal and wind) technologies (Halliday & Ruddell, 2010; Jay & Jeffrey, 2010) and part of the PV value chain (Irvine, 2010) and positive regulatory changes show some encouraging signs already, sub-national industrial and energy policies could aid the transition process for at least four reasons: First, evolutionary economic theory emphasizes the importance of spillovers for niche creation and technological change. Second, regions, cities and localities can take advantage of "first nature" geographies and specialized resources such as firm or research networks. Third, as renewable energy projects need to be implemented at the local scale consent for them is easier to obtain if communities and local authorities are included in project design and realize returns from it. Engagement with localities may thus reduce planning delays and rejections, which constitute one of the major bottlenecks for wind farm and biomass deployment in the UK (BWEA, 2010). Fourth, energy policies have neglected meso-level, decentralized solutions (e.g. district heating) and focused on the international, national and building scales with little regard for the potentially vital role of regions, cities and localities to develop alternative, locally adapted energy scenarios (Scrase *et al.*, 2009; Watson *et al.*, 2010).

Regional Scale

In order to offer an indicator of the regional potential for niche development in the renewable energy sector, Figure 4 maps the total number of patents for the European TL3 regions between 2006 and 2008 (the last year for which data are available). Thus, the information represents current inventive capacity.

Echoing the performance at the national scale, the dominance of Danish regions and a selected number of Spanish and German regions in renewable energy patents is clearly visible, echoing the strength of these countries in the renewable energy sector. Of the top 10 patenting regions, 3 are Danish, 5 are German, 1 is Spanish and 1 is Dutch with Eastern Jutland (DK) producing by far the largest amounts of patents (91.2) followed by Munich (GE) with 56.2 and Südlicher Oberrhein (GE) with 51. While the UK as a whole is among the top 10 of renewable energy patent producers, only Cambridge and Cardiff produced more than 20 patents over the last 3 years. While Cambridge applied

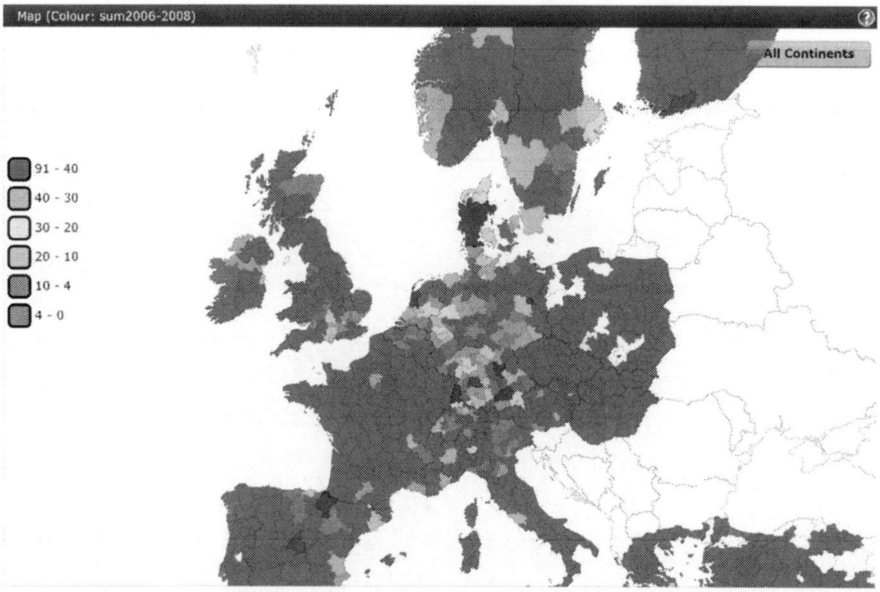

Figure 4. Cumulative number of renewable energy patents (2006–2008) for TL3 regions. *Source*: OECD; mapped with OECD eXplorer.

for patents on a regular basis, the high number for Cardiff is due to a unique high value in 2007. When aggregating all five London TL3s, London has the same patent count as Cambridge (35). Other important regions in the UK are Southampton, Hampshire and Oxfordshire. Once we account for population size (and eliminate regions with less than or equal to six patents in those 3 years) Cambridge, Southampton, Cardiff and the Isle of Wight are all among the top 10, while all German regions but Südlicher Oberrhein drop out (Table 2). Three Danish regions and Navarra remain in the top 10 complemented by Neuchatel (CH). If we calculate the share of renewable energy patents on the total number of patents in order to obtain a relative performance measure, Southampton, Cardiff and the Isle of Wight remain in the top ten together with various North German and Danish regions as well as Navarra and Almeria (ES) (Table 2). While the German, Danish, a number of Spanish and Dutch regions dominate patent counts in RETs in absolute terms, a number of English regions perform well once we control for population size or total patent activity in a region. The Danish regions and Navarra remain among the top in absolute and relative terms, while Northern and some East German regions perform well even after controlling by the total number of patents.

While national policies are probably the main drivers for the development of strong renewable energy sectors and spatial clusters of economic activity, regional industrial policies were employed to attract the PV industry to the depressed economic areas of East GE, Saxony, Thuringia, Saxony-Anhalt and Berlin-Brandenburg explaining the high patent counts in Berlin and Oberes Elbtal/Osterzgebirge (including Dresden) (El-Beyrouty *et al.*, 2009; GTAI, 2009). The dominance of Danish regions (especially Jutland) in the wind power sector was the result of strong deployment incentives combined with carefully targeted industrial policies as well as community ownership of wind farms in the 1980s when the industry took off (Munksgaard & Morthorst, 2008; Cooke, 2010). In the following, the case of Navarra is discussed in greater detail as it demonstrates the importance of incentive policies to increase regional deployment attracting investors and R&D facilities

Table 2. Regional RET patenting activity.

RET Patents per 100,000 inhabitants		Percentage of RET patents on total patents	
TL3 region	RET Patents/ 100,000 pop.	TL3 region	RET Patents/total patents
Vestjylland	38.3	Ostfriesland	25.6
Østjylland	37.6	Navarra	25.0
Southampton	27.5	Vestjylland	19.4
Navarra	23.4	Østjylland	18.7
Cardiff and Vale of Glamorgan	20.6	Schleswig-Holstein Nord	18.2
Isle of Wight	20.3	Almería	18.0
Cambridgeshire CC	19.7	Cardiff and Vale of Glamorgan	17.4
Sydjylland	18.6	Isle of Wight	15.8
Neuchâtel	17.9	Southampton	11.9
Südlicher Oberrhein	16.3	Schleswig-Holstein Mitte	11.2

Source: Data based on OECD Patent Database, http://www.oecd.org/sti/inno/oecdpatentdatabases.htm

to take advantage of spillover and learning effects, which in turn help to accelerate deployment and development of RETs at the national scale.

The development of the wind power cluster (and to a lesser extent the solar industry) in Navarra is the result of resource scarcity in the region, strong government incentives and entrepreneurial talent. As early as 1985, the regional government passed a law to diversify its energy sources, reduce energy dependency and use abundant renewable energy sources, increase energy efficiency and develop new infrastructure for those RETs (Faulin *et al.*, 2006). To support this endeavour, Energia Hidroelectrica the Navarre (EHN) was founded in 1989 by Esteban Morrás supported by capital from the regional government, its dependent savings bank and the regional utility company Iberdrola. The company was set up with the explicit goal to increase the renewable energy share in the region (De Miguel Ichaso, 2000; Fairless, 2007; Gobierno de Navarra, 2010). Confronted with the physical limits to expand the network of small hydroelectric power plants in the region, Morrás turned to Vestas (a Danish company and global leader of wind turbine production) and discussed an ambitious plan for wind power deployment in the region and SP in general. The first turbines were installed in 1994. From 1993 onwards, EHN was in the business primarily to develop aero-generators and wind farm development and in 1994, Gamesa Eólica was founded as a joint venture between EHN, the regional government and Vestas to supply wind turbines to EHN.

The turn to wind as a major power source was supported by the regional government in three ways: First, the government set itself ambitious targets in its 1995/2000 energy plan that it surpassed; Second, in addition to the national FIT, the regional government offered generous investment subsidies to private companies (20% of investment costs between 1994 and 1999 and 30% of investment costs between 2000 and 2004; Faulin *et al.*, 2006) in addition to support for the construction of renewable energy installations (wind farms, solar parks, etc.) totalling €400 million by 2004. Third, local authorities subscribed to the regional plan and public consensus was the result of recognition of the need for wind energy as primary energy source and payments from developers and the local government for the use of land for RET and development of land around installations. As a result, wind capacity was deployed much faster in Navarra than in SP over the period 1995–2004 (Figure 5(a)). Between 1998 and 2000, Navarra's share on SP's installed wind capacity exceeded 20% (while accounting for only 2.5% of total land mass) and in 1996 and 1998 more than a quarter of newly installed capacity was deployed in Navarra (Figure 5(b)).

In addition to providing 65% of total electricity consumption in Navarra, the wind boom attracted a number of key companies to the area including Acciona, the world largest wind farm developer, taking over EHN in 2004, Gamesa Eólica (now among the top three global wind turbine manufacturers), Alstom Power (Ecotechnica) and MTorres (Gobierno de Navarra, 2010). As a result, 14% of the world's aero-generators were produced in Navarra in 2000 (Faulin *et al.*, 2006). Overall, there are more than 100 RET companies employing over 5000 workers in the region. Complementing the strong private sector are the Technological Centre of Renewable Energy Sources that was established in 2002 and the Integrated National Centre for Training in Renewables that was established in 2003, making Pamplona the *de facto* renewable energy centre of SP. Because of its strong institutional support, Navarra was awarded the EU prize for best regional renewable policy in 2004 and attracted considerable attention from political delegations within and outside SP. In this sense, Navarra becomes a showcase for the design of renewable

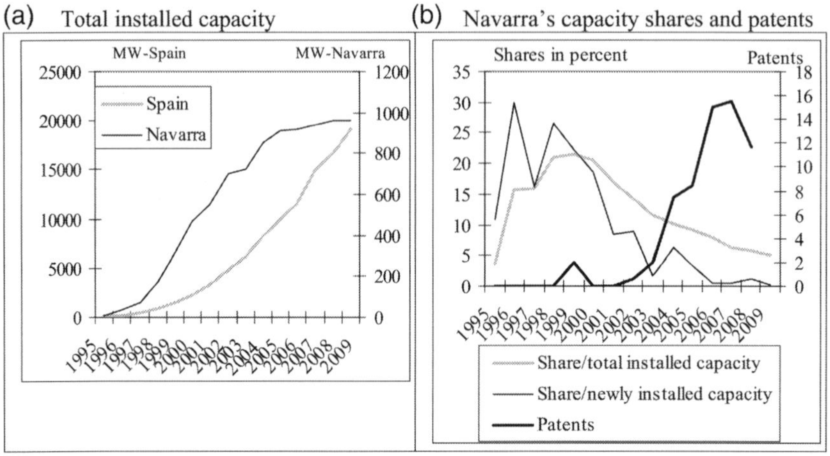

Figure 5. Wind energy deployment in Navarra and SP, 1995–2009.
Source: IDEA (various year): Wind energy in Spain, current status reports.

energy policy that can be reproduced (probably in modified form) somewhere else. While Navarra constitutes only a single case, a number of insights can be gained: First, strong incentive policies at the national scale are necessary for increasing renewable deployment levels. Second, renewable shares can be increased significantly through additional support at the regional scale. Third, deployment leads to R&D, patent activity (Figure 5(b)) and production. Companies and research facilities follow the markets and good institutional support structures (at least in the case of onshore wind since the late 1990s). Fourth, the formation of regional clusters is important as they can become drivers of national competitiveness in the industry. Fifth, strong local incentives and realization of local benefits increase the acceptance of RETs allowing rapid deployment in the first place. Sixth, and complementing the Navarra case study, the regional patent analysis revealed that the relatively successful energy transitions of DK, GE and SP are linked to strong regional patenting activity. For these reasons, attention to the sub-national scale can aid our understanding of energy transitions in general.

Returning to the UK, the large number of patents in Cambridge is not translated into renewable clusters or higher deployment levels in the area or wider UK. The neglect of and need for the development of a green energy technology cluster is explicitly mentioned in Library House's (2007) Cambridge cluster report. However, relatively high patent numbers in Cambridge, London, Oxford and Southampton/Hampshire do indicate inventive activity that could form seeds for regional path creation in RETs. As the case of Navarra and East GE indicated, strong patent performance was led by investment and deployment incentives to translate inventions into viable products and generate regional clusters of specialized economic activity that can then increase national competitiveness in the industry. The UK renewable energy policy tends to be technology-specific and not region-specific, although offshore technologies do generally favour coastal areas with port facilities (DECC, 2011). The following section briefly reviews regional renewable energy policies in the UK and evaluates the potential impact of recent changes in energy and spatial policies to develop industrial clusters and accelerate the UK's energy transition.

Regional Renewable Policies in the UK

While the national government is responsible for promoting technology development and providing deployment incentives, the local level is where new projects are implemented and where formal and informal consent to local projects is given. Situated between the local and the national scale, was the regional scale with government offices (GOs) (set up in 1994), RDAs and regional assemblies (RAs) (set up in 1999), regional stake holders and national organizations with regional operations (such as the Carbon Trust or the Energy Savings trust) charged with devising regional renewable targets, regional spatial strategies favourable to renewable energy, setting up regional R&D centres and energy agencies, and facilitating the development of clusters and supply chains all in the context of increasing regional economic performance. Regions were supposed to operate strategically as they had knowledge of the situation on the ground as well as policy frameworks at the national scale. GOs, RDAs and RAs were the key regional governance bodies devising strategies and implementing them (Smith, 2007; CLG, 2009).

By 2004/2005, GOs administered £9 billion of central government expenditures, employed over 3000 civil servants and coordinated regional renewable energy partnerships (CLG, 2009). RDAs invested £2.2 billion each year into regeneration and were the key source of regional funding for renewable energy. RAs included renewable energy provisions in spatial and housing strategies, providing the key driver for boosting regional capacity in renewable energy (Smith, 2007). The Department of Trade and Industry (DTI) provided each region with £100,000 for facilitating renewable energy governance, partnership building and consultation processes, conducting renewable energy resource assessments and establishing regional targets.[7] Regional targets were developed pragmatically based on regionally available resources, planned housing expansion, businesses already in the region and so forth. While formal governance structures existed, they were endowed with too little financial and administrative support (e.g. regional targets were not binding) to have a major impact on regional renewable deployment in the UK. Other problems to implement renewable energy policy identified by the RAs were (among others):

> Funding uncertainties; inconsistency in obtaining planning permission for wind developments and the way regional planning policies are applied by local authorities; concern over the use of cumulative effects of anecdotal evidence (including unevidenced impact on bird populations) against granting planning permission for further wind farm developments; lack of sufficient local public benefits or securing these consistently to act as an incentive to implementing renewable energy developments and projects; lack of political will and leadership to push through large scale development projects in the face of local opposition. (CLG, 2009, pp. 29–30)

As a result, regional targets were not met. Figure 6 compares deployed renewable capacity in the English regions in 2009 with the 2010 and 2020 regional renewable targets. Judged solely by deployment/target ratios, regional renewable policies have failed. In 2009, only London surpassed its renewable energy target of 2010, while the Southwest and Yorkshire and Humber had installed less than a third of the 2010 target.

While formal regional governance was insufficiently supported, the seed money of the DTI led to the emergence of informal networks. The main tasks were the identification,

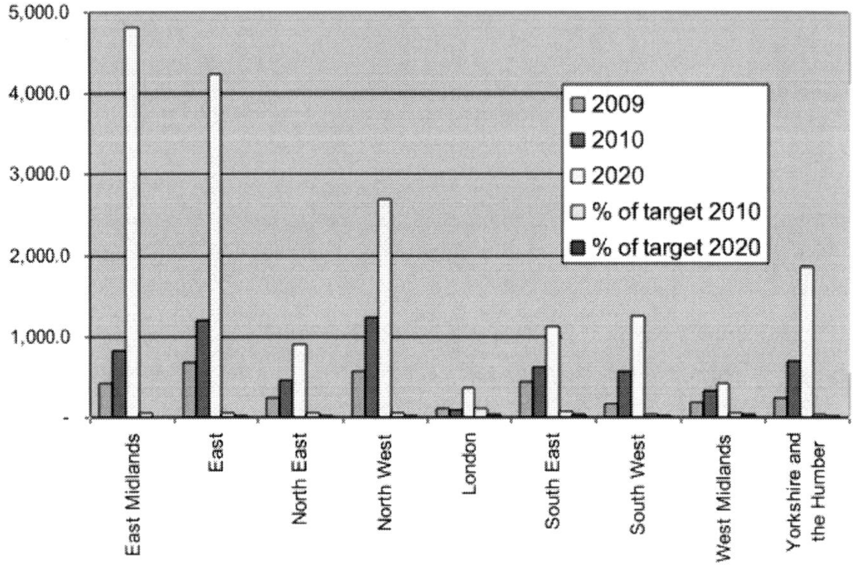

Figure 6. Installed renewable capacity 2009 and target capacities for 2010 and 2020 for nine English regions.

Source: 2009 Capacities: DECC Regional Renewable Statistics; 2010 and 2020 targets from CLG 2009, Table 4.3.

Note: Targets for London, Northeast and West Midlands reflect on-shore targets only.

communication and promotion of the regional energy business potential, the mapping of renewable energy business clusters and support of them, analysis of regional renewable energy supply chains and securing contracts for local firms from entering renewable businesses, creating renewable R&D facilities, assessing skill needs and providing training programmes, and providing tools and services for local businesses to access national and European funding bodies (Smith, 2007). All regions have tried to measure and map existing renewable energy business in their region. For instance, the Northwest RDA estimates that the regional energy and environmental technology sector employs more than 50,000 people (25,000 in the nuclear industry) and over 100 renewable energy companies in the wind and tidal sectors[8] while envirolink Northwest identifies 353 suppliers for renewable energy and 15 research groups in seven universities in the Northwest[9]. Similar exercises took place in all other regions and created a wealth of knowledge and networks with the potential for building stronger renewable energy industries in the regions. Another important step to increasing deployment was regional renewable capacity studies commissioned by DECC (see Note 2) with the intention of regions and local authorities to act on this information, perhaps through development of common, local energy visions around local "first nature" advantages.

The creation of informal networks and formal renewable energy groups (e.g. Southwest region) also resulted in lobbying at the national level in order to help attract international renewable energy capital highlighting the complex relationship between energy policies at the national and regional scales. In the case of the Southwest, the RDA contributed £2 million to a £15 million project to establish a "Wave Hub"off the Cornish coast. In order to enrol DTI support for Wave Hub[10], Southwest region had to organize a national conference on the finance and policy issues surrounding marine technologies. While wave hub was not the only institution pushing for wave technology, the combined power of regional projects and early research outcomes have convinced DECC to continue funding marine energy projects. DECC, the Carbon Trust and BIS now recognize wave and tidal technology as one where the UK has a competitive economic advantage (see also Table 1). The Carbon Trust (2011) estimates a total cumulative market of close to £350 billion peaking in 2040 with £6 billion. As the market leader in marine technologies, the UK should be able to capture a significant part of the potential revenues and the investment to bring the technology to the market seems justified. DECC supports marine energy technology with £20 million over the next 4 years (DECC, 2011) and will establish a marine device testing facility at Narec (the national centre to advance renewable energy) while BIS announced a Technology and Innovation Centre with a focus on offshore renewable energy including offshore wind, wave and tidal technologies.[11] While offshore wind technology has been deployed commercially, wave and tidal technology is still in the development stage with a variety of designs being tested at the moment to maintain diversity and avoid lock-in to inferior technological trajectories (Arthur, 1994; Stirling, 2011). In addition to wave and tidal technologies, offshore wind power requires excellent port facilities, expertise in installation of offshore wind towers and offshore grid connection where the UK does have a strong competitive advantage through its expertise in North Sea oil and gas drilling (Jay & Jeffrey, 2010). The UK also benefits from "first-nature" advantages in these technologies, making it attractive to foreign investors to take advantage of those resources. All three eastern regions – Northeast, Yorkshire and Humber, East of England – develop strategies to take advantage of the coming "offshore boom".

According to the CLG (2009), RDAs, RAs and GOs were among the key players in developing and implementing national renewable policies through embedding renewable energy targets in regional spatial strategies and regional economic development plans. However, regional strategies and RDAs will be abolished and local authorities and Local Enterprise Partnerships (LEPs) are supposed to take over. In a letter[12] to the chief planning officers (dated 6 July 2010) revoking the regional spatial strategies, the chief planner stated that

> Through their local plans, authorities should contribute to the move to a low carbon economy, cut greenhouse gas emissions, help secure more renewable and low carbon energy to meet national targets, and to adapt to the impacts arising from climate change. In doing so, planning authorities may find it useful to draw on data that was collected by the Regional Local Authority Leaders' Boards (which will be made available) and more recent work, including assessments of the potential for renewable and low carbon energy.

The capacity studies commissioned by the regions do contain information at the level of local authorities that could be used to design local energy visions around existing resources. Inclusion and support is easier to obtain at the local level if people, businesses and policy makers can rally around common goals and engage in the design and implementation of local and transition pathways that gel with local endowments and result in benefits for local communities through job creation and revenues from energy sales (Scrase *et al.*, 2009). In response to calls for exploitation of meso-level renewable sources (Watson *et al.*, 2010) and the need to incentivize renewable deployment for local communities, new legislation allows local authorities to develop their own renewable energy sources and keep the profits for reinvestment in their communities (Huhne, 2010). A sense of ownership by communities may also reduce the amount of planning objections and speed up the planning process for renewable energy sources identified as major problem by the RDAs. This process could be supported by devolving national targets to local authorities coupled with stronger enforcement of non-compliance and incentives to meet or exceed them. The Draft National Planning Policy Framework (Department for Communities and Local Government 2011) appears to favour renewable energy in the planning process but renewable policy still takes a backseat to regeneration and economic development and it is unclear how the new LEPs will be able to use informal renewable networks and infrastructure as well as information on local businesses compiled by the RDAs to move the transition process forward or whether the abolishment of RDAs and regional spatial strategies may result in a planning and administrative vacuum that is filled by powerful lobbying groups.

Conclusion

Climate change and energy security are the main drivers of the transition process towards a low-carbon socio-technical energy regime. There are a number of possible ways to achieve this transition but in the long run, renewable energy generation is imperative. The UK is characterized by a highly centralized system where efficiency gains (low-cost energy) were the main goal for private producers, grid operators and suppliers. In order to unlock the existing energy system, a number of changes must occur simultaneously. First, the comparative analysis at the national scale together with results from econometric analysis (Johnstone *et al.*, 2010) demonstrate that strong government incentives for demand creation (such as FITs) and technology push (public R&D investment) are necessary to speed up deployment and development of RETs. Manipulation of landscapes in which energy regimes are embedded is essential. Second, the UK has an R&D, innovation and skills gap in most renewable technology areas that needs to be closed rapidly to take advantage of the "green revolution" and thus support niche development. The recent increase in R&D budgets and patenting activity supplemented by support or RETs by DECC, BIS and NPS for Energy indicate steps in the right directions but may be funded insufficiently for the energy sector to meet its targets (it is estimated that £100 billion need to be invested between now and 2020 to meet the targets set out in the roadmap). Third, in order to speed up niche innovations, a multi-scalar approach to energy transitions should be applied with sub-national policies complementing national policies. Regional and local policies can take advantage of locally specific regime configurations, benefit from localized spillovers, help maintain technological diversity and are more likely to enrol a large number of actors in the transition project, triggering the development of common regional visions that will be easier to implement than "distant" national guidelines. The recent change in legislation to allow local authorities to own and benefit economically from renewable energy projects and regional capacity studies for local authorities to identify locally specific renewable energy policies are potentially important steps in that direction. However, radical changes to planning and energy policies could generate uncertainty and further delay the development and deployment of renewable energy. It needs to be seen whether LEPs will be able to work with local authorities on developing sustainable, local growth strategies built on the creation and deployment of renewable energy sources. As the case of Navarra demonstrated, strong regional support for the deployment of RETs can then result in cluster formation that can improve national comparative advantages in the sector. While the sub-national scale is important to speed up RET deployment and niche development and garner local support for RETs, they cannot function independently from strong incentive policies at the national scale. While it is important to consider the sub-national scale in transition research and policy, an exclusive focus on the sub-national scale and neglect of national and supra-scale processes will result in an incomplete understanding of those processes.

Notes

1. http://cordis.europa.eu/funen/spotlight-07_en.html (accessed 15 September 2011).
2. The reports can be found at http://www.decc.gov.uk/en/content/cms/meeting_energy/renewable_ener/ored_news/ored_news/method_assess/method_assess.aspx (accessed 15 September 2011).
3. Japan, the USA and Korea are also major players in innovation and production of renewable energy sources, but in order to control for the impact of EU policies, the paper restricts the comparative analysis to EU countries.
4. http://epp.eurostat.ec.europa.eu/portal/page/portal/energy/data/database (accessed 15 August 2011).
5. http://stats.oecd.org/index.aspx (accessed 18 June 2011).
6. TL3 correspond with NUTS-3 regions in most countries (see Maraut et al., 2008 for more details). RETs include wind, solar thermal, solar PV, geothermal, marine and wave/tidal technologies.
7. Although there are other investment subsidies, the figure does not compare favourably with the €136 million invested in Navarra (a region of 500,000 people) between 1995 and 2004 (Fairless, 2007) or the $1.2 billion of subsidies provided by the German government for solar PV firms in East GE until 2008 (GTAI, 2009).
8. http://www.nwda.co.uk/areas-of-work/supporting-business/key-sector-support/energy–environment.aspx (accessed 15 September 2011).
9. http://www.envirolinknorthwest.co.uk/ (accessed 15 September 2011).
10. "Wave hub is a marine testing facility whose first customer is Ocean Power Technologies and is complemented by the Peninsula Research Institute for Marine Renewable Energy, a centre of excellence delivering world-leading research, facilities and technology transfer in marine energy, excellent port infrastructure and an established supply chain in the South West of England" (www.wavehub.co.uk).
11. www.innovateuk.org/deliveringinnovaiton/technology-and-innovation-centres/offshore-renewable-energy.ashx (accessed 15 September 2011).
12. http://www.communities.gov.uk/documents/planningandbuilding/pdf/1631904.pdf

References

Alhemade, F., Frenken, K., Hekkert, M. & Schwoon, M. (2009) A complex systems methodology to transition management, *Journal of Evolutionary Economics*, 19(4), pp. 527–543.

Arthur, B. (1994) *Increasing Returns and Path Dependence in the Economy* (Ann Arbor, MI: The University of Michigan Press).

Boschma, R. (2009) *Evolutionary Economic Geography and its Implications for Regional Innovation Policy*. Report for the OECD. Version 20 February 2009. Available at http://econ.geo.uu.nl/boschma/OECD.pdf (accessed 19 March 2012).

Boschma, R. & Van der Knaap, G. A. (1997) New technology and windows of locational opportunity: Indeterminacy, creativity and chance, in: J. Reijnders (Ed.) *Economics and Evolution*, pp. 171–202 (Cheltenham, UK: Edward Elgar).

Broeckner, W. S. (2007) CO_2 arthimethic, *Science*, 315, p. 1371.

BWEA. (2010) *England's Regional Renewable Energy Targets*. Progress Report. London: BWEA.

Carbon Trust. (2011) *Marine Renewables Green Growth Paper* (London: Carbon Trust).

CLG. (2009) *Renewable Energy Capacity in Regional Spatial Strategies*. Final Report. London: Department for Local Communities and Government.
Coenen, L. & Diaz Lopez, F. J. (2010) Comparing systems approaches to innovation and technological change for sustainable and competitive economies: An exploratie study into conceptual commonalities, differences and complementaries, *Journal of Cleaner Production*, 18(12), pp. 1149–1160.
Cooke, P. (2010) Regional innovation systems: Development opportunities from the "green turn", *Technology analysis and Strategic Management*, 22(7), pp. 831–844.
De Miguel Ichaso, A. (2000) Wind power development in Spain, the model of Navarra, *DEWI Magazin Nr. 17*. August 2000.
Department for Communities and Local Government. (2011) *Draft National Planning Policy Framework*. London: Department for Communities and Local Government. Available at http://www.communities.gov.uk/documents/planningandbuilding/pdf/1951811.pdf (accessed 19 March 2012).
Department of Energy and Climate Change (DECC). (2010) *Annual Energy Statement*. DECC Departmental Memorandum. July 27, 2010. London: DECC.
Department of Energy and Climate Change (DECC). (2011) *Renewable Energy Road Map* (London: DECC).
Dosi, G. (1982) Technological paradigms and technological trajectories: A suggested interpretation of the determinants and directions of technical change, *Research Policy*, 11(3), pp. 147–162.
Edquist, C. (1997) *Systems of Innovation. Technologies, Institutions and Organizations* (London: Pinter).
El-Beyrouty, K., Meimanalieve, A-S, Petrosyan, L. & Singh, D. (2009) *Germany's Photovoltaic Cluster. PED 329* (Cambridge, MA: Harvard University Press).
Essletzbichler, J. (2012) Evolutionary economic geographies, in: T. Barnes, J. Peck, & E. Sheppard (Eds) *The Wiley-Blackwell Companion of Economic Geography*, pp. 179–194 (London: Wiley-Blackwell).
Essletzbichler, J. & Rigby, D. (2007) Exploring evolutionary economic geographies, *Journal of Economic Geography*, 7(6), pp. 549–571.
Fairless, D. (2007) Energy go around, *Nature*, 447(28), pp. 1046–1048.
Faulin, J., Lera, F., Pintor, J. M. & Garcia, J. (2006) The outlook for renewable energy in Navarre: An economic profiles, *Energy Policy*, 34(15), pp. 2201–2216.
Foxon, T. J., Hammond, G. P. & Pearson, P. J. G. (2010) Developing transition pathways for a low carbon electricity system in the UK, *Technological Forecasting and Social Change*, 7(8), pp. 1203–1213.
Freeman, C. & Perez, C. (1988) Structural crises of adjustment, business cycles and investment behaviour, in: G. Dosi, C. Freeman, R. Nelson, G. Silberberg & L. Soete (Eds) *Technical Change and Economic Theory*, pp. 38–66 (London: Pinter).
Geels, F. (2002) Technological transitions as evolutionary reconfiguration processes: A multi-level perspective and a case-study, *Research Policy*, 31(8–9), pp. 1257–1274.
Geels, F. (2005) *Technological Transitions, a Co-Evolutionary and Socio-Technical Analysis* (Cheltenham: Edward Elgar).
Geels, F. & Schot, J. (2007). Typology of socio-technical transition pathways, *Research Policy*, 36(3), pp. 399–417.
Gobierno de Navarra (2010) *2010 Horizon. Navarra and Renewable Energy Sources* (Navarra: Gobierno de Navarra).
Grabher, G. (1993) The weakness of strong ties—The lock-in of regional development in the Ruhr Area, in: G. Grabher (Ed.) *In The Embedded Firm*, pp. 255–277 (London: Routledge).
GTAI. (2009) *The Photovoltaic Industry in Germany—The World's Strongest PV Cluster*. Germany Trade and Invest: February 2009. Berlin: GTAI.
Haas, R., Panzer, C., Resch, G., Ragwitz, M., Reece, G. & Held, A. (2011) A historical review of promotion strategies for electricity from renewable energy sources in EU countries, *Renewable and Sustainable Energy Reviews*, 15(5), pp. 1003–1034.
Halliday, J. & Ruddell, A. (2010) *UKERC Energy Research Landscape—Wind Energy*. UKERC. Available at http://ukerc.rl.ac.uk/Landscapes/Wind.pdf (accessed 15 September 2011).
Hascic, I., Johnstone, N., Watson, F. & Kaminker, C. (2010) *Climate Policy and Technological Innovation and Transfer: An Overview of Trends and Recent Empirical Results* OECD Environment Working Paper No. 30. Paris: OECD.
Hekkert, M. & Negro, S. (2009) Functions of innovation systems as a framework to understand sustainable technological change: Empirical evidence for earlier claims, *Technological Forecasting & Social Change*, 76(4), pp. 584–594.

HM Government. (2009) *The UK Low Carbon Transiton Plan*, (National Strategy for Climate and Energy), Norwich: TSO.
Hodson, M. & Marvin, S. (2010) Can cities shape socio-technical transitions and how would we know if they were? *Research Policy*, 39(4), pp. 477–485.
Hodson, M., Marvin, S. & Hewitson, A. (2008) Constructing a typology of H2 in cities and regions—An international review, *International Journal of Hydrogen Energy*, 33(6), pp. 1619–1629.
Hughes, T. P. (1983) *Networks of Power: Electrification in Western Society, 1880–1930* (Baltimore, MD: Johns Hopkins University Press).
Huhne, C. (2010). *Unlocking Local Power*. Department of Energy and Climate Change. Available at www.decc.gov.uk/assets/decc/News/376-unlocking-local-power-huhne-letter.pdf (accessed 19 March 2012).
Innovas. (2009) *Low Carbon and Environmental Goods and Services: An Industry Analysis*. Report commissioned by the Department of Business and Regulatory Reform (BERR). London: BERR.
Instituto para la Diversificatcion Ahorro de la Energia (IDAE) (2010) *Renewable Energies in Spain* (Madrid: IDEA).
International Energy Agency (2011) *Clean Energy Progress Report* (Paris: IEA).
Irvine, S. J. C. (2010) *UKERC Research Landscape—Solar energy*. UKERC. Available at http://ukerc.rl.ac.uk/Landscapes/Solar.pdf (accessed 15 September 2011).
Jacobsson, S. & Bergek, A. (2011) Innovation system analyses and sustainability transitions: Contributions and suggestions for research, *Environmental Innovation and Societal Transitions*, 1(1), pp. 41–57.
Jaffe, A. B., Trajtenberg, M. & Henderson, R. (1993) Geographic localization of knowledge spillovers as evidenced by patent citations, *Quarterly Journal of Economics*, 108(3), pp. 577–598.
Jamasb, T., Nuttall, W. J. & Pollitt, M. (2008) The case for a new energy research, development and promotion policy for the UK, *Energy Policy*, 36(12), pp. 4610–4614.
Jay, B. & Jeffrey, H. (2010) *UKERC Research Atlas Topic—Wave*. UKERC. Available at http://ukerc.rl.ac.uk/Landscapes/Marine.pdf (accessed 15 September 2011).
Johnstone, N., Hašcic, I. & Popp, D. (2010) Renewable energy policies and technological innovation: Evidence based on patent counts, *Environ Resource Econ*, 45(1), pp. 133–155.
Kemp, R. (1994) Technology and the transition to environmental sustainability—The problem of technological regime shifts, *Futures*, 26(10), pp. 1023–1046.
Klaasen, G., Miketa, A., Larsen, K. & Sundqvist, T. (2003) *Public R&D and Innovation: The Case of Wind Energy in Denmark, Germany and the United Kingdom*. Interim Report IR-03-011. Laxenburg: Institute for Advanced Studies.
Klepper, S. (2002) The capabilities of new firms and the evolution of the U.S. automobile industry, *Industrial and Corporate Change*, 11(4), pp. 645–666.
Library House. (2007) *Looking Inwards, Reaching Outwards. The Cambridge Cluster Report—2007* (Cambridge: The Library House).
Lundvall, B. A. (1992) *National Systems of Innovation: Towards a Theory of Innovation and Interactive Learning* (London: Pinter).
Maraut, S., Dernis, H., Webb, C., Spiezia, V. & Guellec, D. (2008) *The OECD REGPAT Database: A Presetnation*. OECD Science, Technology and Industry Working Papers, 2008/2. Paris: OECD. doi: 10.1787/241437144144.
Martin, R. (2010) Rethinking regional path dependence: Beyond lock-in to evolution, *Economic Geography*, 86(1), pp. 1–27.
Martin, R. & Sunley, P. (2006) Path dependence and regional economic evolution, *Journal of Economic Geography*, 6(4), pp. 395–437.
Messner, D., Schellnhuber, J., Rahmstorf, S. & Klingenfeld, D. (2010) The budget approach: A framework for a global transformation toward a low-carbon economy, *Journal of Renewable and Sustainable Energy*, 2, 031003. http://dx.doi.org/10.1063/1.3318695.
Munksgaard, J. & Morthors, P.E. (2008). Wind power in the Danish liberalised power market – Policy measures, price impact and investor incentives. *Energy Policy*, 36(10), pp. 3940–3947.
Neffke, F., Henning, M. & Boschma, R. (2011) How do regions diversify over time? Industry relatedness the development of new growth paths in regions, *Economic Geography*, 87(3), pp. 237–265.
Nelson, R. (Ed) (1993) *National Innovation Systems* (Oxford: Oxford University Press).
Nelson, R. (2001) The coevolution of technology and institutions as the driver of economic growth, in: J. Foster & S. Metcalfe (Eds) *Frontiers in Evolutionary Economics*, pp. 19–30 (Cheltenham: Edward Elgar).
Nelson, R. & Winter, S. (1977) In search of a useful theory of innovation, *Research Policy*, 5(1), pp. 36–76.

Norgaard, R. B. (1994) *Development Betrayed: The End of Progress and a Coevolutionary Revisioning of the Future* (London: Routledge).

OECD. (2009) *Patent Statistics Manual* (Paris: OECD).

REN21. (2011) *Renewables 2011*. Global Status Report. Paris: REN21 Secretariat.

Rip, A. & Kemp, R. (1998) Technological change, in: S. Rayner & E. Malone (Eds) *Human Choice and Climate Change,* vol. 2, pp. 327–399 (Columbus, OH: Batelle Press).

Scott, A. J. (1988) *Metropolis: From the Division of Labor to Urban Form* (Berkeley, CA: University of California Press).

Scrase, I., Stirling, A., Geels, F. W., Smith, A. & Van Zwanenberg, P. (2009) *Transformative Innovation,* A report to the Department for Environment, Food and Rural Affairs, SPRU—Science and Technology Policy Research, University of Sussex.

Shove, E. & Walker, G. (2007) CAUTION! Transitions ahead: Politics, practice and sustainable transition management, *Environment and Planning A*, 39(4), pp. 763–770.

Smith, A. (2007) Emerging in between: The multi-level governance of renewable energy in the english regions, *Energy Policy*, 35(12), pp. 6266–6280.

Smith, A., Stirling, A. & Berkhout, F. (2005) The governance of sustainable sociotechnical transitions, *Research Policy*, 34(10), pp. 1491–1510.

Smith, A., Voss, J. P. & Grin, J. (2010) Innovation studies and sustainability transitions: The allure of the multilevel perspective and its challenges, *Research Policy*, 39(4), pp. 435–448.

Späth, P. & Rohracher, H. (2010) "Energy regions": The transformative power of regional discourses on sociotechnical futures, *Research Policy*, 39(4), pp. 449–458.

Stirling, A. (2011) Pluralising progress: From integrative transitions to transformative diversity, *Environmental Innovation and Societal Transitions*, 1(1), pp. 82–88.

Storper, M. & Walker, R. (1989) *The Capitalist Imperative* (Malden, MA: Blackwall).

Unruh, G. C. (2000) Understanding carbon lock-in, *Energy Policy*, 28(12), pp. 817–830.

Van den Bergh, J. C. J. M., Truffer, B. & Kallis, G. (2011) Environmental innovation and societal transitions: Introduction and overview, *Environmental Innovaiton and Societal Transitions*, 1(1), pp. 1–23.

Verbong, G. & Geels, F. (2007) The ongoing energy transition: Lessons from a sociotechnical, multi-level analysis of the Dutch electricity system (1960–2004), *Energy Policy*, 35(2), pp. 1025–1037.

Walker, G. (2008) What are the barriers and incentives for community-owned means of energy production and use? *Energy Policy*, 36(12), pp. 4401–4405.

Walker, G. & Devine-Wright, P. (2008) Community renewable energy: What should it mean? *Energy Policy*, 36(2), pp. 497–500.

Watson, J., Scrase, I. & Stapleton, L. (2010) *Transforming the UK's Energy System—Policies for the 2020 Renewables Target and Beyond.* A report for Friends of the Earth. Sussex: University of Sussex, Sussex Energy Group, SPRU.

Transversality and Transition: Green Innovation and New Regional Path Creation

PHILIP COOKE*,**

*Centre for Advanced Studies, Cardiff University, Cardiff, Wales, UK, **School of Development Studies, Aalborg University, Aalborg, Denmark

ABSTRACT *Since Paul David first published his economic histories of path-dependent innovation, the subject has exerted fascination upon scholars of innovation and technological change and, latterly, regional scientists and economic geographers. This paper speaks of the third and fourth of these communities in the main, though it may have theoretical and empirical elements of interest to the first two as well. It begins with an overview of recent perspectives and critiques concerning the relevance of the path dependence concept to the understanding of regional economic development and its associated governance. It then goes on to discuss the contribution of evolutionary economic geography to thinking about "branching" from path dependence and the creation of new paths. Evidence for key generic spatial processes of path transition is provided before the main content of the paper concludes with new insights into the contributions of regional innovation policy to path evolution. Conclusions are then drawn.*

1. Introduction

Path dependence is treated as an important concept in economic history, innovation studies and socio-political history and by evolutionary economic geographers. This is clearly due to attempts made to conceptualize why "history matters" rather than simply supplying narrative accounts of its effects. Like historians, geographers have repeatedly felt the need to proclaim that "geography matters" too, and the recourse to theory has been important in underlining that assertion as much as theory-building has been in some aspects of history (David, 1985). By far the most serious effort has been made by the evolutionary economic geography/regional science community (Cooke & Morgan, 1998; Cooke, 2005; Martin & Sunley, 2006; Belussi & Sedita, 2009; Martin, 2010). Absent from this list are Storper and Walker (1989) not because their book had no relevance to thinking about path dependence, written as it was from a Marxist perspective, but rather because it has, with its notion of "windows of locational opportunity", had perhaps too much. This is evidenced in a recent regional evolutionary economics volume on "cluster emergence" (Fornahl *et al.*, 2010) where this concept is of key importance in 7 of the 13 contributions. It will be argued in the concept review section of this paper that heavy reliance on such a passive category detracts from rather than contributing significantly to evolutionary science.

This paper then goes on to advocate a more agency-centred perspective drawing theoretical inspiration from the fields of evolutionary urban economics (Jacobs, 1969) and evolutionary innovation economics (Schumpeter, 1975); both Jacobs and Schumpeter have lately been shown to be important architects of a powerful theoretical framework capable of explaining regional evolution. The two master concepts or axes of this frame are the Jacobian concept of "diversity" or its evolutionary variant of "variety", especially (after Frenken *et al.*, 2007) "related variety", and Schumpeter's concept of "regional innovation" as rendered in Andersen's (1994, 2007, 2011) magisterial reconstruction and synthesis of Schumpeter's complete political economy (see also Andersen, 2002, 2009). Together, these give a clear insight into the dynamics of regional evolution. As we shall see, from the concept of Jacobs, explanations of both firm and, more importantly, cluster mutations by means of cross-pollination of knowledge (knowledge spillovers) leading to innovation by means of geographically proximate inter-cluster "collisions" of ideas can be deduced. From that of Schumpeter, we get the mechanism for these mutations for, as Andersen (2011) shows, against many unreflective conflations of "innovator" and "entrepreneur", they are, for Schumpeter, rigidly distinctive categories. The former "recombines" knowledge, while the latter acts upon and "commercializes" it. These are conceptually distinct skillsets, even though they may occasionally be combined empirically in a single actor. Accordingly, imitative "swarming" (or clustering) is principally an "entrepreneurial event". By these combined process mechanisms, regional evolution, including "branching" to new path creation, occurs. In this respect, this contribution transcends the problem of path creation stated by Garud and Karnøe (2001) in their review, which assigns virtual omniscience to "entrepreneurship" and no role for the separate capability of the Schumpeterian "innovator" in innovation and change.

In what follows, the family of concepts, such as "path dependence", "windows of opportunity" and "chance effects", associated with the orthodoxy will be briefly reviewed and critiqued and the newer, evolutionary elements such as "related variety", "mutation" and "innovation transversality" (on the latter, see Cooke, 2011a; Harmaakorpi *et al.*, 2011) critically reviewed. This will be a prelude to an extended exemplification of key elements of the Jacobs–Schumpeter framework by comparison with a handful of illustrative regional cases. Importantly, these mix process mechanisms emanating primarily from market transactions, on the one hand, and innovation policy governance interventions, on the other hand. Finally, a concluding section brings the argument of the paper to fruition.

2. Concepts of Consequence: A Review

It is clear from the preceding section that this perspective on "new path creation" presumes that this occurs in reality. There can be no doubt that it does, although the nature, extent and key mechanisms of change, on the one hand, but possibly more interesting in such a context, "arrested development" and (regional) economic stasis or decline, on the other hand, require far more embellishments than are currently on display (see also Martin, 2011). This is an extraordinarily important issue since almost nowhere in the evolutionary economics or economic geography literature are these creative processes explored. However, the following offers some assistance, though their implications will have to be pursued elsewhere due to space limitations. In his book on complexity theory, Kauffman (2008) introduces two invaluable concepts applicable to innovation in human and natural systems. The first is "preadaptation" (also known in ecology as "exaptation"), whereby a pre-existing idea or artefact can be applied sufficiently out of context to constitute a "cognitive reversal" to produce an innovation. It is a clear case of a Schumpeterian knowledge recombination as a source of innovation. A vivid illustration occurs in the history of tractor technology. Early designs incorporated a necessary heavy engine which, nevertheless, broke successive chassis models. However, pervasive innovation ensued when Ford head engineer Eugene Farkas successfully designed the engine block, transmission and axle housings bolted together to form the basic, chassis-free, structure of the Fordson F tractor in 1916. The second important concept is the "adjacent possible" (on this, see also Johnson, 2010), which is a cumulative capacity in which the more variety the system displays, "the easier is the creation of still further novelty" (Kauffman, 2008, p. 151). However, because further out from the present, human capability for prediction decays dramatically, such novel moves are generally fairly of short range but adjacent.

The relationship of these two concepts designed to capture different dimensions of the "recombinative" character of innovation to path dependence and, crucially, path *inter-dependence* appear in Figure 1. The core concepts underlying new path creation are shown in the figure. This is relatively straightforwardly explained in the following. The oval in the top half of the diagram represents elements in the historic economic development trajectory of a region. The region contains industries some of which display both path

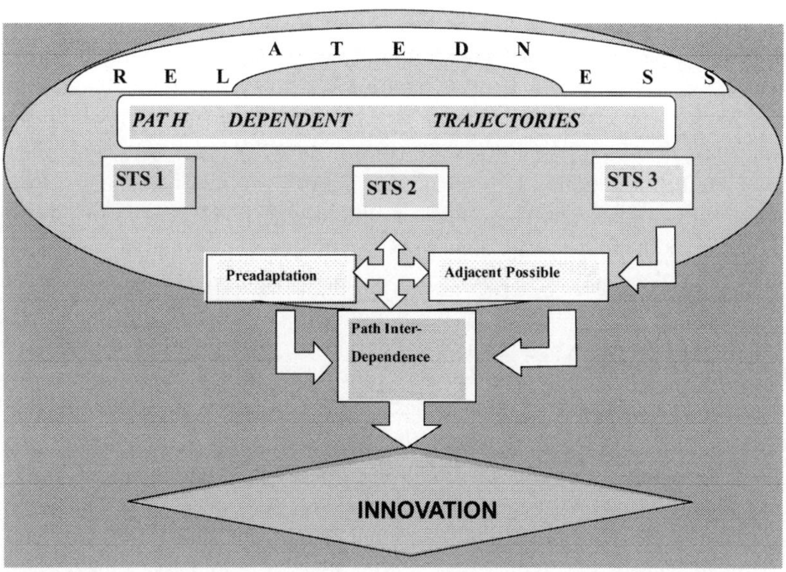

Figure 1. Path inter-dependence and regional innovation. *Source*: Centre for advanced studies.

dependence and relatedness. Path dependence represents the manner in which their technological paradigms and their socio-political regimes have evolved in the region. Hence, these are social constructs rather than narrow technologies or industrial products and processes. In evolutionary complexity theory after Kauffman (1995), they represent centres of energy or *clusters* that may "self-organize" to interact. This is the key dynamic force animating change in complex adaptive systems. It begins informally around such "socio-technical systems" (STSs) through "cafeteria effects", "gossip" and "knowledge spillovers" among different cluster members. Where clusters are isolated, this does not occur and the regional economic profile remains static. Contrariwise, regional innovation and change are products of evolutionary development, whereby it is noted that

> ... The more diverse the economic web, the easier is the creation of still further novelty ...[leading to]... a positive correlation between economic diversity and growth. (Kauffman, 2008, pp. 151–160)

Relatedness is expressed in the particular "attractors" that recombine innovative knowledge. It is broader than the evolutionary economic geography idea of "related variety" (Frenken *et al.*, 2007) which tends to limit itself—unnecessarily from an innovation point of view—to neighbouring or otherwise similar industries or skillsets. Much if not most of the innovation results from unpredictable knowledge recombinations (Arthur, 2009), accordingly innovative relatedness can only be understood *ex post*. Elsewhere, even "related variety" proponents have accepted this in the notion of "revealed related variety" (Neffke & Svensson-Henning, 2008; Boschma & Frenken, 2011). What evolutionary economic geography lacks is a concept of the creative mechanism by which path inter-dependent innovation actually occurs. This is supplied by the complexity science concepts described above of "preadaptation" and the "adjacent possible" which bring about path inter-dependent knowledge recombinations among STSs as shown in Figure 1.

Now, it remains to be described what, where and how "transversality"—one of the key concepts of this paper—relates to the foregoing analysis. If we return to the regional paradigm" and "regional regime" distinction mentioned above, it is relatively easy to see that the first refers to economic development processes, while the second refers to regional governance. The drivers of the former are markets and the latter networks. Clearly, just as it is possible to have policy failure caused by, for example, inadequate lateral information flow among administrative "silos" at the regime level, it is even more the case in relation to market failure associated with "asymmetric information" (Akerlof, 1970), competition, confidentiality and intellectual property rights blockages at the paradigm level. Accordingly, "transversality" is the social agency dimension of the rather passive notion of "relatedness" including "related variety". Transversality is invoked in two major ways: first, by firms realizing that market opportunity may lie in knowledge interaction across sectoral or cluster boundaries by their own initiative and, second, by an intermediary regime organization facilitating regional or other knowledge interactions among incumbent firms. An example of the first kind is when agro-food firms interact with automotive firms in projects to develop novel biofuels, while the second is illustrated by the German regional development agency that arranges innovation "theatre" which displays novelties from one industry or cluster that are applicable in wholly different ones (Cooke *et al.*, 2010). Hence, this is a clear vindication from the study of human

Regional Innovation	Variety: Low	Variety: High
High	• *Creative Destruction* • *Low Path Dependence* • *Regime Shift* • *Innovation 'Push'* • *E. G. O'vik, Sweden*	• *Mutation/Transversality* • *Early Path Dependence* • *Demand-driven Innovation* • *Co-evolutionary Transition* • *E.G. Jutland, DK*
Low	• *Equilibrium* • *High Path Dependence* • *Branching* • *User-driven Innovation* • *Incremental Change* • *E.G. Värmland, SE*	• *Punctuated Equilibrium* • *Late Path Dependence* • *Renewal* • *Design-driven Innovation* • *Market Paradigm Shift* • *E.G. Lombardy*

Figure 2. Varieties of path-dependent economic developments.

and natural adaptive systems of two core evolutionary economic geography concepts of "relatedness" and "path dependence" in circumstances where "proximity" (Boschma, 2005; Boschma & Frenken, 2006) facilitates the promotion of "transversality" among related industries or clusters.

Such theory and evidence of active social agency invoking "emergence" of innovation tends to undermine the so-called explanation of change in social phenomena as represented in the notion of "window(s) of locational opportunity" (WLO; see also Cooke, 2010a). This is because as authors such as Martin and Sunley (2006, 2010) have argued, the WLO perspective lacks agency, except *ex post*, and assumes that which must be explained and the principals responsible for change.[1] Equally, their critique of traditional notions of path dependence as being somewhat negative, neoclassical and equilibrium oriented vitiates this as a dynamic explanation for regional growth. Their notion of path inter-dependence shows a greater promise for utilizing path dependence as a subsidiary conceptual element in a more satisfying overarching theoretical perspective consistent with evolutionary economic geography, which itself is uncomfortable with equilibrium assumptions except in the short term. As Figure 1 reveals, it also gets us closer to an understanding of new path creation. For its proactive connotations towards social agency, we refer to the general approach as "transversality", embodying both "relatedness" and "revealed related variety" (Boschma & Frenken, 2006). The discourse already denotes a certain ecosystem-like set of economic presences and possibilities in a given regional economy. For simplicity, it assumes geographical proximity circumscribing relational proximity, consistent with the concepts of preadaptation and the adjacent possible.

2.1 *A Jacobian–Schumpeterian Matrix*

We have thus seen how regional economic branching is more clearly capable of being rationally explained by virtue of the *transversality* perspective than the "window of opportunity" or traditional "path dependence" and "chance" approaches. As noted, they really have no theory of change, rather "path dependence" is a thesis about continuity and the others leave it in the lap of the gods. Moreover, when put to work in the regional economic sphere, even path dependence loses much of the relevance it may have had once for accounting for curiosities, like the survival of "QWERTY" on the iPad. Perhaps what follows is a little harsh, but the fact that opera is still noticeably path dependent on the Italian language or that the names for most Japanese industrial equipment essentially remain German because that was the language of the instructing engineers of the 1868 Meiji Restoration is roughly as important to regional economic development.

The next step in this exegesis is to harness the work of Jacobs and Schumpeter to sustain the discourse on relatedness and transversality. We start with that of Jacobs (1969) where we find what have subsequently been termed "Jacobs' externalities" to be a useful concept (e.g. Boschma & Wenting, 2007; Boschma & Ledder, 2010), but with variable and unsuspected constraints. In the latter chapter, Dutch banking spinoffs are shown to have less impact than spinoffs have on manufacturing, whether as in Klepper's (2002) pioneering study of spinoffs and clustering in Detroit car manufacturing or as in Boschma and Wenting's (2007) similar study in the UK. This is because Dutch banks had variable survival rates and spinoffs from existing banks only came later rather than at the cluster's beginning.[2] It is concluded that in the end "…. Amsterdam was just lucky to have many start-ups with pre-entry experience in the banking and related sectors" (Boschma & Ledder, 2010, p. 206). Why the above-average share of start-ups (compared with early spinoffs)? It is cultural, and probably, though this is to surmise, due to the Netherlands' late industrialization, because

> … entry levels remained low until the 1890s. One of the reasons was that it was considered a sign of weakness to [borrow] money from a bank. In the 1890s, this resistance for credits from banks started to disappear … and …[became] normal among a growing number of entrepreneurs. (Boschma & Ledder, 2010, p. 198)

Accordingly, the nature of Jacobs' externalities continues to warrant further and deeper research as a consequence of study that moves away from a manufacturing to a service template. Having said this, Wenting's (2009) study of the *haute couture* fashion clusters in Paris, London, New York and Milan finds spinoffs and labour mobility to be the principal causes of growth in each, but particularly in the largest and oldest, namely Paris. So, some urbanization economies (for that is what Jacobs' externalities define) derive externality benefits from user-driven demand (banking), while others, notably cars and fashion, derive theirs more from design-driven demand. In the latter case, designers "propose" products to the market using their expertise in symbolic knowledge manipulation, and users (consumers) take it or leave it. Famously, in the case of the Ford Edsel, they left it.

Jacobs' externalities or knowledge spillovers can thus be user- or design-driven in their innovation demand. In Verganti's (2006, 2009, Figure 2) discussion of these, he holds that design-driven innovation is more *radical* than the user driven one, which, following Dosi (1982), he sees as always incremental, thus path dependent by nature.

By contrast, Dosi (1982) sees "technology push" innovation as capable of provoking regime and/or paradigm change in dominant technologies and innovation trajectories. This raises issues of time and depth in different types of radical innovation. This now seems long-term and pervasive (epochal) or short-term and focused (episodic) (Cooke, 2011a). Verganti's (2006) field of interest is the Lombardy regional innovation system, particularly its overlapping and interacting design-intensive furniture and kitchenware clusters, where his equivalent to a technological paradigm is a socio-cultural paradigm. Instead of technology, its discourse is *meaning*, and by changing meanings, designers, like technologists, are capable of changing paradigms and regimes through innovations that entrepreneurs commercialize. This innovator–entrepreneur division is quite pronounced here, with "circles" of external as well as internal designer-innovators iterating rounds of "meaning analysis" to set the tone for new ranges of design-intensive products to be "proposed" to consumer markets (Pisano & Verganti, 2008).

Accordingly, this intersects with Schumpeter, who, as Andersen (2011) shows, was rather rigid about the separate functions of *innovator* as recombiner of knowledge (old and new) and *entrepreneur* as commercializer and imitator, responsible for cluster "swarming" of the innovation. This is important because it recognizes the "related variety" of specialized functions operating in the process of introducing innovation into the prevailing socio-technical "landscape". Andersen (2011) demonstrates this by means of an imaginary dialogue with Schumpeter. This focuses on Schumpeter's fifth and most radical form of innovation, represented by the introduction of railways to new lands ("railroadization" as he calls it),

> .... We ask him for answers that can be derived from his basic evolutionary theory as well as for his major examples:

- Can a Schumpeterian innovation be implemented in incremental steps? No. It involves a jump away from old routines. Example? The replacement of the mail coach by the railroad.
- Can the innovation be implemented by its inventor? No. It is carried out by an entrepreneur based on already existing inventions. Example? The use of largely pre-existing technology for the first successful general-purpose railroad: the Liverpool–Manchester line from 1830.
- Can the innovation be induced by consumers? No. It has to be made by an entrepreneur and to be enforced upon consumers. Example? Consumers initially had to be persuaded to use the railroads.
- Can the innovation be implemented within the framework of an already existing firm? No. It presupposes the creation of a new firm. Example? A mail coach firm never moved into the railroad business.
- Can the innovation be implemented with the saved money of an entrepreneur? No. This entrepreneur needs to borrow money for his innovative project. Example? The financing of the railroad projects .... (Andersen, 2011, pp. 30–31)

This is to be contrasted with Garud and Karnøe's (2001) reification of entrepreneurship as the key change agent. From a Schumpeterian perspective, the entrepreneur is the "imitation" agent not responsible for novelty or innovation. At extremes, as Dosi (1982) sees it, certain innovations with radical qualities could be the handmaiden of those "gales of creative destruction" by means of which old path dependences and their associated regimes could be terminated, or at least severely weakened, and new ones set in motion. Here, Jacobs makes a brief, but important re-entry based on variety, relatedness and the transversality that may arise in contexts of abundant knowledge spillovers and high lateral absorptive capacity. This is the point at which an innovation mutates through the interaction of these forces as entrepreneurs act upon the knowledge symbiosis effected by inter-cluster, inter-firm or inter-individual innovators. From this, a new cluster or sector may emerge to prominence, as discussed above, there branching, or in a more Schumpeterian context of "creative destruction" *breaking* with the former regional path dependence. Much depends on the definition of "radical" in innovation studies, a subject returned to in Section 4. In Figure 2, we simply refer to this as "regional innovation" since this term captures the two sides of the Schumpeterian innovation idea: on the one hand, "railroadization" or the opening up of new developmental pathways and, on the other hand, "creative destruction" or the undermining of old developmental pathways. This rather dramatic theoretical bifurcation is moderated usefully by interaction with Jacobs' concept of variety. Both these dimensions are then given the designations "high" or "low", and from the resulting two-by-two matrix, we draw out exemplars of distinctive forms and degrees of developmental path dependence, which are then explicated in the accounts that follow. In passing, while perceived as the "purest" form of radical innovation, "railroadization" need not be understood as unproblematic, let alone perfect. The lawlessness, violence and genocide that characterized "how the West was won" bear witness to the costs of massive social upheaval. As is shown elsewhere (Cooke 2010c), Iceland's dash from fish to finance, was also highly fraught, though fortuitously, less so than on the American Frontier.

3. Varieties of Regional Path Dependence

In what follows, only two accounts will be provided of key variables related to Figure 2. This is because only two of them refer to renewable energy. Each refers to varieties of path dependence and the roles of related variety and distinctive innovation drivers in contexts such as the creation of new developmental pathways, incremental path-dependent modernization and relative regional economic equilibrium. In this paper, because of the interest in new path creation, we do not consider cases of regional economic stasis or decline; however, it can be hypothesized that most would fall in the bottom left-hand corner of Figure 2, that is, low-variety and low regional innovation.

3.1 *North Jutland, Denmark: A Local Green Platform in Global Energy Markets*[4]

Denmark's (Figure 3) high related variety axis intersecting a moderate regional innovation axis offers a relatively smooth portrayal of regional economic modernization. Nevertheless, it still reveals change of a potentially *transitional* magnitude (re: climate change) occurring within certain inherited, path-dependent industrial parameters. Some explication of key terms is warranted here. It can be recalled from Section 2 that "transversality" is the condition whereby a region contains a number of industries or clusters that display relatedness and from which innovations that cross-pollinate between such distributed knowledge bases may be created. In a context where such innovations are both produced and

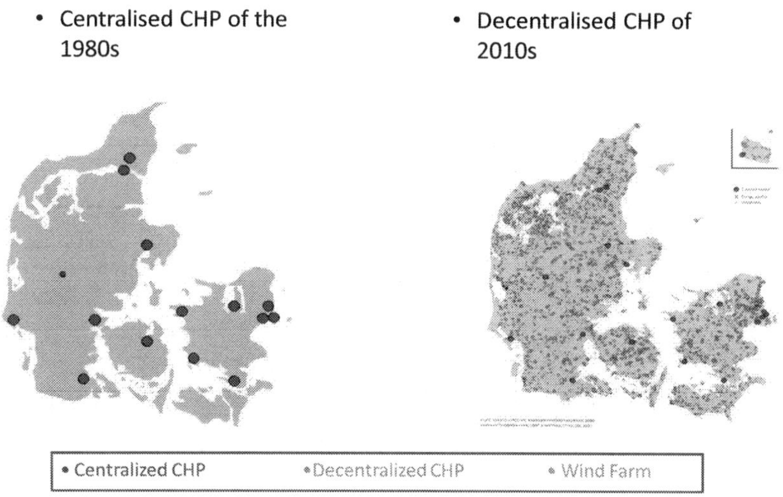

Figure 3. Denmark's shift to decentralized CHP. *Source*: Danish energy authority.

consumed, the region in question may be engaged in an innovative *transition* involving at least some degree of technological regime change (Freeman & Perez, 1988; Cooke, 2010b). Accordingly, it may warrant the appellation of "transition region", which because of its relatively pervasive adoption of innovation in related production and consumption profiles (e.g. renewables) acts as a kind of a "living laboratory" of the future to which experts and policy-makers from other regions and countries may be drawn to assess its transferability. A particularly precise form of such transition occurs in relation to eco-innovation which has been proposed as "co-evolutionary transition" by leading "system innovation" authors (Kemp, 2002; Geels, 2004, 2006) focusing on innovative responses to anthropogenic climate change. Since this is empirically shown to occur in a spatially uneven manner, the designation "transition regions" is proposed for those meso-units breaking out of path dependence upon fossil fuels in production and consumption (Cooke, 2011b).

In the account that follows, it can be seen that "transversality" is a crucial feature of "co-evolutionary transition" to decarbonization[5] because many technologies and institutions must combine systemically through "strategic niche management" to achieve it. The complexity of this tends to mean that eco-innovations are incremental though rapid in their overall effect. This means that they may both embody "preadaptations" and explore the "adjacent possible" simultaneously or sequentially. An instance of preadaptation in the evolution of Jutland's wind turbine technology lies in the history of its world-leading renewable energy company *Vestas*. This firm began life in agricultural engineering, specifically in the manufacturing of milk-cooler technology for the regional dairy industry. It later utilized this fan-based fluid-stirring technology to produce turbo-coolers for marine engines. Accordingly, *Vestas* was obviously attuned to evolving new market niches for its core competence in rotary blade technologies. It was also facing stiff competition from neighbouring competitors such as Sweden's global agro-engineering firm *Alfa-Laval*. The upshot of this is that *Vestas* began to take advantage of the Danish government's "strategic niche management" of wind turbine technology as represented in a subsidy to consumers to install early wind turbines in the same dairy farms *Vestas* had hitherto serviced with milk-coolers. An instance of successful exploration of the adjacent possible, where unlike in preadaptation, no previous technological base that can be applied to a new use exists, occurs in relation to the integration of innovations that come to replace a hitherto dominant technological paradigm and regime. As discussed in more detail below and shown in Figure 4, North Jutland has developed a world-leading, innovative renewable energy district heating system platform. Until the 1980s, this was of limited scale because energy was supplied to the region by three large coal-fired combined heat and power (CHP) stations. Once again, the Danish government, enforcing popular will to develop renewable energy, established an incentive scheme that led to a major transition in energy provision towards renewables throughout the country (Figure 3). As renewable district heating evolved into CHP, new challenges arose because unlike fossil fuels, which are continuously available, renewable energies are not. Accordingly, innovation became focused upon how to manage transfers of renewable power sources affordably in the power station. This led to combined wind and solar energy solutions and, later, the quest for new forms of renewable energies whose regional and local availability would keep the costs down. This has led to the development of biogas energy which recycles animal waste from the regional dairy industry as an added energy source in the mix. This is a good illustration of exploration of the adjacent possible or "proximity spillovers" occasioned by the necessity for innovations to facilitate new path creation, in this case from a carbonized energy to a renewable energy STS. This shows very clearly how regime change in demand may ultimately induce a paradigm-transcending, post-hydrocarbon production/consumption "landscape". The latter is a long-term expectation in "co-evolutionary transition" theory, capable of being glimpsed in "transition regions" which, nevertheless, must co-exist with the broader, global, hydrocarbon path dependence and its prevailing regulatory regime.

The present account is important not least because it draws attention to the role of *discourse articulation* in stimulating change from path-dependent development.[6] Critique can be a powerful trigger of social change, rather as neoliberal discourse formed the basis for a thoroughgoing critique of the path-dependent model of state ownership of assets, so in Denmark and elsewhere, two critical discourses were articulated to undermine

- 'Innovative Region: Flexible District Heating' Platform
- Biogas, Biomass, Solar Thermal, Wind - 'plug-ins'
- 'Social Network' >100 'system' & 'solution' firms
- Aalborg U, Municipalities, DTI, VåkstForum Fund (40 mn.DK bid).
- 'Aggregators' or 'system integrators' include:
- Arcon Solar (Velux VHK), Xergi, Logstor (Pipework), Baracon (Biogas), Grundfos
- Humvel, NIRAS, EnergiPlan (consultants)

Figure 4. North Jutland's green regional innovation system. *Source*: Centre for advanced studies.

prevailing path dependence circa the 1960s. The first of these was an anti-nuclear energy discourse; the second was its obverse, a pro-renewable energy discourse. In combination and through protest-based social movements, this discourse brought about a reversal of government policy, the suspension of nuclear energy policy and the redirection of Denmark's nuclear research (the Risø laboratory at Roskilde) towards renewable energy research (Jørgensen & Karnøe, 1995; Karnøe & Jørgensen, 1996). Thus, knowledge from research conducted on modern wind turbine prototypes was transferred to regions where industrial expertise and weather conditions were perceived as supportive of creating early path dependence on renewable fuels.

A further important point is that in such demand-driven innovation settings, subsidies are a necessary element of "strategic niche management" and are found being made to consumption rather than only to production. The Danish case involved what *ex post* is revealed as a successful consumer subsidy policy regime that more than paid for itself in tax returns from wind turbine production. From the early 1970s, government subsidies were made available to users of first-generation wind turbines. This sustained the industry, initially based largely upon domestic demand, and enabled the north and mid-Jutland-based cluster to out-compete its main rivals in California and others in Europe. The user subsidy stimulated experimentation, knowledge spillovers and niche market evolution in regionally "path-dependent" trajectories in both Jutland and California. But the Reagan administration jettisoned wind energy subsidies, while in Denmark they continued until a right-wing coalition entered government in 2000. By that time, the Danish design had evolved considerably from its path-dependent roots in agricultural and marine engineering where *Vestas* began. Meanwhile, the Californian design atrophied around its inspiration, propeller-driven aircraft. Already something of an anachronism, the two-blade, pointed downwind turbine design proved inferior to the three-blade, pointed upwind Danish solution. *Vestas*, Denmark's national champion, now has 40% of the world wind turbine market and has been joined in its Aarhus–Aalborg cluster by the likes of Germany's *Siemens*, which acquired the other main Danish companies, *Suzlon* from India and *Gamesa* from Spain. Including home market production of turbines in Germany and Spain, these European producers, along with Denmark, have 70% of world turbine production capacity with employment of 133,000 and global demand far from saturated.

However, this was only the beginning of the exploitation of path-dependent relatedness in new ways. Markets had opened up for renewably fuelled CHP stations for district heating. Demanding customers are the municipalities, most of whom run local energy supply companies, and about 60% of Denmark's citizens rely upon it. Municipalities seek a balanced supply and order customized mixes of a variety of related energies involving biomass, biogas, wind, solar and marine energy depending on location and the type of solution required. Enormous export markets for district heating have opened up in mature and emerging markets faced with climate change and "peak oil" constraints, notably India and China. Within North Jutland (Figure 4) is a community of some 100–150 specialist renewable energy firms that combine to meet this demand. An example of incremental innovation is that of the pipework firm *Logstor*, now a North Jutland District Heating company having branched from marine to power station technology and innovated a pre-insulated dual pipe system to minimize heat loss by fitting the cold water input pipe inside the hot water pipe. Together, the district heating firms, municipalities, university laboratories and technology transfer agencies created an association entitled *Innovative Region: Flexible District Heating* with characteristics described in Figure 4.

This echoes the 2007 regionalization of Denmark's administration into five, one of which is North Jutland. It constitutes both a regional innovation system and a "transition region" designation because, on the one hand, of its interactive commercialization and knowledge generation sub-systems and, on the other hand, its regional production/consumption of renewable energy. The former consists of networks of firms in supply chains focused around the district heating engineering platform while belonging to distinctive renewable energy business segments. These are, nevertheless, capable of being system integrated by lead "aggregator" firms such as solar thermal specialist *Arcon*, biogas contractor *Xergi*, green engineering firm *Grundfos* or consultants *NIRAS* into consortia for plant assembly. Supporting this sub-system is a knowledge enterprise support subsystem consisting of public laboratories, regional development agency, municipalities and technical agencies such as the Danish Technological Institute. In 2008, the consortium made a bid to the regionally administered Danish Growth Fund (*Väkstfonden*) for "user-driven design and innovation" (Ministry of Foreign Affairs of Denmark, 2008). A €5 million award was forthcoming to the consortium, now known as *Flexenergi*, which currently has five innovative district heating pilot projects underway in the region.

Briefly to reprise, this is an illustrative case of high-variety cluster mutation involving embedded marine and agricultural engineering firms branching into the energy industry exploiting key design transversalities among wind turbine blade technology, pipework, gearing, energy damping and so on that evolved into design, prototyping and production for a new industry. Accordingly, the region was a beneficiary of early path dependence (discussed by Martin, 2010, after Page, 2006) and, following the vicissitudes of knowledge transfer, foreign competition and consumer subsidy now operate in a world of global demand-driven, mostly incremental, innovation for its renewable energy products and services. As stressed in the account, the importance of co-evolution among industry, institutions and innovation discourses and practices was crucial to this achievement.

3.2 *Slow, Incremental Innovation from Limited, but Evolved Relatedness: Processum biorefinery*

The second exemplar derived from the varieties of regional path dependence analyses and taxonomies presented in Figure 2 is the peripheral Swedish region of Norrland and its green epicentre at Örnsköldsvik (otherwise O'vik) historically path dependent on forest products, notably the pulp and paper industry. However, the interest for this study lies in the relatively constrained but undoubtedly impressive "branching" that has been induced by transversality policy in an otherwise fairly unprepossessing regional setting. This is an innovation platform management model that focuses upon a single high-variety capability, represented in the case of the *Processum* biorefinery hub at Örnsköldsvik in Norrland, Angermanland province. Its innovative applications are relevant to Sweden's Grand Challenges of Energy, Sustainability and Healthcare, but include textiles, paints and food ingredients. Centred upon a former pulp and paper mill, this geographically confined platform occupies the *Processum* business park. In its proximity are various process industry facilities as shown in Figure 5, including biochemicals, biofuels such as biomass, bioethanol, biodiesel and biogas, packaging, pharmaceuticals, substitute cotton, sugar, food, nutritional ingredients and construction materials. These act as raw

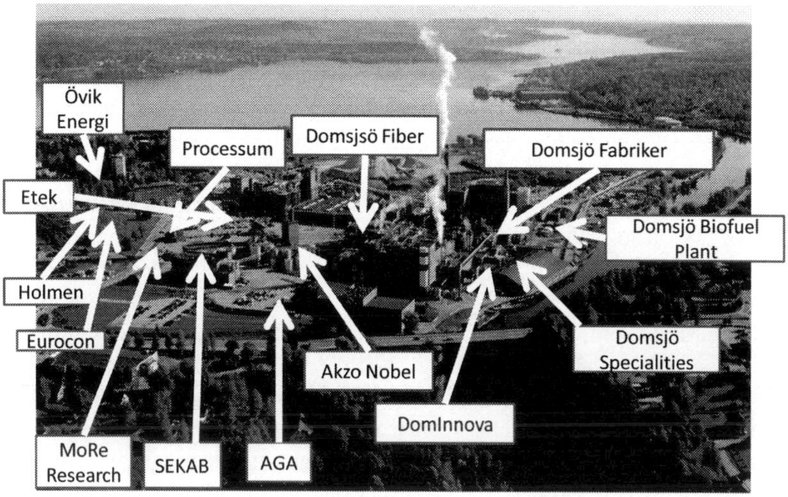

Figure 5. The Örnsköldsvik processum innovation platform.

material feedstocks to other industries, the range of which is broad and includes paint, manufactured by cluster-member Akzo Nobel, and organic cotton substitute, for which interactions are evolving through Västra Gotäland's "Smart Textiles" cluster at Boräs near Gothenburg with the H&M (Hennes & Mauritz) clothing retail company. The reason for this innovative translation of specialist cellulose "comfort fibre" into cotton textiles is that today an era of 'peak cotton' prevails in the textile industry because of the unsustainable production methods and steeply rising price of cotton on the world market. Simultaneously, sugar from forest products is currently mixed with imported sugarcane to produce bioethanol, contributing to reductions in sugarcane imports. Biomedical applications of forest products include tissues and cloths used in surgery, bio-"scaffolds" that support body parts temporarily before dissolving and, again working with Smart Textiles, woven blood vessel replacements.

As noted, Sweden has developed a "Grand Challenges" national innovation strategy, following the EU initiative of the same name. The *Processum* biorefinery addresses the Swedish Grand Challenges of Energy, Sustainable Production and Healthcare. As implied above, the centrepiece of this complex is the once obsolescent *Modo* sulphite mill, which was long ago transformed into a (failing) pulp and paper mill. After 4 years on the closure list, it was acquired by a private consortium in 2000. They set a new course and turned the ailing pulp mill into a highly profitable biorefinery. Under the ownership of *Domsjö*, the plant was earning SEK 3 millions per day, producing from wood cellulose organic raw cotton substitute. Its other recycled pulp waste products include biodiesel and bioethanol with construction planned of a new plant to produce biogas from pulp waste.

Today's global "peak cotton" situation means that this organic cotton product sold at \$800–\$1000 per tonne in 2008 but at \$1000–\$5000 per tonne in 2011 with 85% being exported to China. On 19 April 2011, the biorefinery was acquired by Indian spinning company *Aditya Birla* at a cost of \$415 million. They realized that more value added could be extracted if the basic pulp mill was transformed into an organic cotton mill.

Major investment in new spinning and weaving technology was seen by *Domsjö* as a necessity before enabling any automated weaving products to be produced competitively in Sweden. Part of the *Birla* investment will be invested in such capabilities in low labour cost India. A pilot plant for assessing yarn-spinning capabilities was under consideration by *Domsjö* before the acquisition. Europe's first pulp mill-biorefinery had been achieving a 40% return on investment, at a time when pulp and paper mills in general were struggling to break even. After the crisis of 2008, when the global pulp and paper industry plunged into a loss of $8 billion, the top 10 in both Europe and Canada remained in the red, although there were signs of recovery in the US by 2011. The now acquired *Domsjö Fabriker* mill was to continue sales of fabric to existing medical textile customers as part of the deal while preparing to extend its biorefining capacity with the installation of the Chemrec biogasifier, expected to start up in 2013. It would convert the spent brown liquor of the sulphite process into 40 million gallons a year of automotive fuel, namely BioDME and BioMethanol. Processum also makes construction materials such as ligno-sulphonate, which is dried and used in concrete, making it stronger and with less water content, and cellulose derivatives as thickeners for construction plaster, while green liquor sludge is tested for dirt-road binding where it acts as an innovative dust inhibitor material. The links with Smart Textiles and the GöteborgBIO cluster initiative include supplying inputs to health industry customers such as Mölnlycke Medical Fabrics AB, a world-leading producer of single-use surgical and wound products where weaving of tissue and surgical cloth has been implemented at Gothenburg. A project on blood vessel replacement textiles is also under way with Smart Textiles at Boräs, near Gothenburg. With GöteborgBIO, verification of bio-"scaffolds" bionics products is a further subject of ongoing project discussions. Elsewhere in the O'vik complex, SEKAB (Figure 5) plans a new torrefaction unit to extract three green ethanol products from black liquor to be used in varieties of gasification. Overall, Processum, as a Swedish platform initiative, funds 50–60 projects, most of which are taken to at least the innovation assessment stage.

Hence this is a model based on initial low-variety but high 'preadaptation' amd 'adjacent possible' innovation. The institutional setup that has arisen from modest policy action and initiatives facilitates evolution of new pathways albeit relying ultimately on the regional core capabilities in pulp and paper production. In other words, the regional economy is relatively equilibriated and strongly path dependent on forestry and its associated support industries, demonstrating evolutionary branching (e.g. towards organic textiles, biofuels and green construction inputs). The regional innovation support system is highly focused on green user-driven innovation, which, according to Dosi (1982), is fundamentally incremental compared with the more radical "technology push" or "design push" kind. Accordingly, what may appear to be new paths can relatively be easily traced back to initial path dependence upon forestry in *Processum*.

4. Conclusions

The main message of this analysis is that possessing or building variety in regional economies is generally a better regional development strategy than seeking to specialize, especially anew, in unrelated activity. In this respect, Jacobs is a superior regional policy guide to Porter. Even the low-early variety, relatively low innovation Norrland region rose among Swedish regions as it evolved its moderately diversified platform.

There are three further conclusions arising from the conceptual and empirical accounts in the foregoing analysis. First, this paper demonstrates the importance of the concept of "innovation" to any discussion of path dependence and the emergence or creation of new paths. In order to do that, however, the thrust of the paper involved re-considering certain shibboleths of innovation system studies. Among these was the "new orthodoxy" (circa Lundvall, 1988; Von Hippel, 1988, 2005) that users and producers combined vertically for innovation in ways overlooked entirely by an earlier generation of production-focused innovation analysts. However, this assumption has remained relatively unexplored in the intervening decades until recently. Accordingly, it is clear that attention must be devoted to time, industry and horizontal proximity in thinking about predominating but also changing forms of innovation impulse. As an exemplar, attention was drawn to design-driven innovation, which is claimed to be paradigm transcending although it is clearly necessary to distinguish "epochal" from "episodic" kinds of innovation radicalism to clarify the adaptive use of core innovation concepts, such as technological and socio-cultural regimes and paradigms.

Second, these were related to a discussion of key concepts in thinking about path dependence and the creation of new paths. These included purported mechanisms by which change in regional economic trajectories was said to occur—largely by chance from "windows of locational opportunity". It was argued that this approach offers very little by way of explanation for important processes such as cluster emergence (Fornahl *et al.*, 2010). Path dependence was argued to be clearly a dependent variable, the importance of which to regional change may be, on the one hand, over-estimated and yet, as the empirical sections of the paper showed, more or less useful on occasions when harnessed to a key change mechanism, notably innovation and its varieties. Evolutionary economic geography's key concept of relatedness was assessed and found to be the most important independent variable in regional economic evolution. Here lies the origin of path interdependence (Martin & Sunley, 2010) energized by "preadaptation" and the "adjacent possible" after Kauffman (2008). This is because together they furnish an explanation for "branching" or indeed "jumping the rails" from path dependence (Örnskoldvik's new paths in textiles, biofuels, etc. may look radical, but all are basically path-dependent upon wood products). Nevertheless, these are outstanding examples of the power of transversality to exploit relatedness to penetrate apparently unrelated market spaces. Relatedness and its cognate, transversality, which embraces regional variety and conscious policy action to exploit it, were shown to be important elements in any explanation of regional change. In North Jutland, the new regional agency has stimulated innovation in district heating systems, while in Norrland, Sweden, Processum has stimulated the policy to support regional innovation platforms supported with innovative projects' novel outcomes and indeed pathways.

Third, and linked to the insights of Jacobs on variety and Schumpeter on regional innovation, a taxonomy was drawn up representing different degrees of path-dependent changes. These ranged from the radical shift away from historic trajectories implied by high innovation and low variety, transition early but mutating co-evolutionary and transversal model in which demand-driven innovation contributed to co-evolutionary transition, to the punctuated evolution/late path dependence of the design-driven innovation model with episodically radical change implications for market collectivities and the equilibriated, incremental, highly path-dependent model in which user-driven innovation predominates (Figure 2; the Värmland exemplar). This taxonomy was then stress-tested against empirical reality by adducing exemplars from Denmark and Sweden. A high level of isomorphism was shown to connect the taxonomy to each test case. This transforms the taxonomy into a typology which can be tested by future research. The results support the insights of evolutionary economic geography rather against the over-equilibriated assumptions of classic path dependence formulations.

Acknowledgements

This paper was prepared for the "New Path Creation" workshop at Trinity College, Oxford, 5–7 September 2010. The author thanks the organizers and participants for their valuable comments. The author also thanks Koen Frenken of the University of Eindhoven for advice on the history of "related variety" and Ron Boschma of the University of Utrecht for his constructive and informative comments on other theoretical aspects of the paper.

Notes

1. It can be objected that Storper and Walker allow for agency: explicitly that human agency will make the difference in some regional contexts, while not in others (acknowledgement to Ron Boschma for pointing this out). But this is rather ambiguous. Does it mean that the same agency works in location A but not in location B? Or that success in location A requires little or no agency? There clearly have to be vision, agency and resources to move through the WLO. The "winners" thus gain their fortune from Pasteur's "prepared mind". The "losers" did not prepare or do it well enough. The WLO thus seems to be a post-rationalization and is, in itself, just a "medium" rather than an active social agent. Maybe it is simply a metaphor, but not necessarily a helpful one that also invites "chance" "explanation". Incidentally, Sydow et al. (2009) show that efforts at chance explanations are in effect lazy since these authors utilize painstaking historical inquiry to reveal social agency explaining all cases cited. This criticism can also be extended to "dartboard" theory as promulgated by the likes of Ellison and Glaeser (1997).
2. Ron Boschma comments that the study did not investigate the Amsterdam banking cluster from its very beginning (but only from 1850 due to data availability), so this citation does not tell the whole story of the paper.
3. Readers interested in the other case material, including accounting for Iceland's transition from fish to finance as a radical branching into a new developmental pathway and Lombardy's socio-cultural regime changing breaks with previous fashion path dependences, may refer to Cooke (2010c).
4. Both the ensuing accounts are based on original documentary and interview research in the regions in question. The Danish research has been ongoing alongside my adjunct professorship at Aalborg University since 2006 and conducted with the help of my colleague there, Søren Kerndrup. The Swedish research was conducted in a project for the Swedish Economic Development Agency (*Tilväxtverket*) in January 2010.
5. The concept of co-evolution, however, is not limited to this field. Compare, for example, Nelson (1994), Nelson (2008) and Cantwell *et al.* (2010). However, these authors are not concerned with related variety or specific industries. Rather they concentrate principally upon the co-evolution of technology markets and regulatory regimes. I am grateful to Martin Heidenreich for these reminders.
6. Originating in Laclau and Mouffe's (1985) theorization of regime evolution and hegemony, it is illuminatingly deployed in Davenport and Leitch's (2009) account of pro- and anti-genetically modified organism discourse strategies in agricultural politics.

References

Akerlof, G. (1970) The market for "lemons": Qualitative uncertainty and the market mechanism, *Quarterly Journal of Economics*, 84, pp. 488–500.

Andersen, E. (1994) *Evolutionary Economics: Post-Schumpeterian Contributions* (London: Pinter).

Andersen, E. (2002) Railroadization as Schumpeter's standard example of capitalist evolution: An evolutionary-ecological account, *Industry and Innovation*, 9, pp. 41–78.

Andersen, E. (2007) *Schumpeter's Evolution* (Aalborg: Aalborg University Press).

Andersen, E. (2009) *Schumpeter's Evolutionary Economics: A Theoretical, Historical and Statistical Analysis of the Engine of Capitalism* (London: Anthem).

Andersen, E. (2011) Schumpeter and regional innovation, in: P. Cooke, B. Asheim, R. Boschma, R. Martin, D. Schwartz & F. Tödtling (Eds) *The Handbook of Regional Innovation & Growth* (Cheltenham: Edward Elgar).

Arthur, B. (2009) *The Nature of Technology* (London: Penguin).

Belussi, F. & Sedita, S. (2009) Life cycle vs. multiple path dependency in industrial districts, *European Planning Studies*, 17, pp. 505–528.

Boschma, R. (2005) Proximity and innovation. A critical assessment, *Regional Studies*, 39, pp. 61–74.

Boschma, R. & Frenken, K. (2006) Why is economic geography not an evolutionary science? Towards an evolutionary economic geography, *Journal of Economic Geography*, 6, pp. 273–302.

Boschma, R. & Frenken, K. (2011) Technological relatedness and regional branching, in: H. Bathelt, M. Feldman & D. Kogler (Eds) *Beyond Territory: Dynamic Geographies of Knowledge Creation and Innovation* (London: Routledge).

Boschma, R. & Ledder, F. (2010) The evolution of the banking cluster in Amsterdam, 1850–1993, in: D. Fornahl, S. Henn & M. Menzel (Eds) *Emerging Clusters* (Cheltenham: Edward Elgar).

Boschma, R. & Wenting, R. (2007) The spatial evolution of the British automobile industry, *Industrial and Corporate Change*, 16(2), pp. 213–238.

Cantwell, J., Dunning, J. & Lundan, S. (2010) An evolutionary approach to understanding international business activity: The co-evolution of MNEs and the institutional environment, *Journal of International Business Studies*, 41, pp. 567–586.

Cooke, P. (2005) Regional transformation and regional disequilibrium: New knowledge economies and their discontents, in: G. Fuchs & P. Shapira (Eds) *Rethinking Regional Innovation & Change: Path Dependency or Regional Breakthrough?* (Berlin: Springer).

Cooke, P. (2010a) Jacobian cluster emergence: Wider insights from "green innovation" convergence on a Schumpeterian "failure", in: D. Fornahl, S. Henn & M. Menzel (Eds) *Emerging Clusters* (Cheltenham: Edward Elgar).

Cooke, P. (2010b) Matrix policy: Rationales and good examples, in: A. Eriksson (Ed.) *The Matrix: Post-cluster Innovation Policy* (Stockholm: Vinnova).

Cooke, P. (2010c) Transversality and transition: Branching to new regional path dependence (2010), *Papers in Evolutionary Economic Geography* 10:10, University of Utrecht, Urban & Regional Research Centre.

Cooke, P. (2011a) Transversality and regional innovation platforms, in: P. Cooke, B. Asheim, R. Boschma, R. Martin, D. Schwartz & F. Tödtling (Eds) *The Handbook of Regional Innovation & Growth* (Cheltenham: Edward Elgar).

Cooke, P. (2011b) Transition Regions: Regional-national eco-innovation systems & strategies, *Progress in Planning*, 76(3), pp. 105–146.

Cooke, P. & Morgan, K. (1998) *The Associational Economy* (Oxford: Oxford University Press).

Cooke, P., De Laurentis, C., MacNeill, S. & Collinge, C. (2010) *Platforms of Innovation* (Cheltenham: Edward Elgar).

Davenport, S. & Leitch, S. (2009) Creating space for the successor: The discourse strategies of pro- and anti-GM factions regarding the future of agriculture in New Zealand, *European Planning Studies*, 17, pp. 943–962.

Dosi, G. (1982) Technological paradigms and technological trajectories: A suggested interpretation of the determinants and directions of technical change, *Research Policy*, 11, pp. 147–162.

Ellison, G., & Glaeser, E.L. (1997) Geographic Concentration in U.S. Manufacturing Industries: A Dartboard Approach, *Journal of Political Economy*, 105(5), pp. 889–927.

Fornahl, D., Henn, S. & Menzel, M. (Eds.) (2010) *Emerging Clusters* (Cheltenham: Edward Elgar).

Freeman, C. & Perez, C. (1988) Structural crisis of adjustment, business cycles and investment behaviour, in: G. Dosi *et al.* (Eds) *Technical Change & Economic Theory* (London: Pinter).

Frenken, K., van Oort, F. & Verburg, T. (2007) Related variety, unrelated variety and regional economic growth, *Regional Studies*, 41, pp. 685–697.

Garud, R. & Karnøe, P. (2001) Path creation as a process of mindful deviation, in: R. Garud & P. Karnøe (Eds) *Path Dependence and Creation*, pp. 1–38 (London: Lawrence Erlbaum).

Geels, F. (2004) From sectoral systems of innovation to socio-technical systems: Insights about dynamics and change from sociology and institutional theory, *Research Policy*, 33, pp. 897–920.
Geels, F. (2006) Co-evolutionary and multi-level dynamics in transitions: The transformation of aviation systems and the shift from propeller to turbojet (1930–1970), *Technovation*, 26, pp. 999–1016.
Harmaakorpi, V., Tura, T. & Melkas, H. (2011) Regional innovation platforms, in: P. Cooke, B. Asheim, R. Boschma, R. Martin, D. Schwartz & F. Tödtling (Eds) *The Handbook of Regional Innovation & Growth* (Cheltenham: Edward Elgar).
Jacobs, J. (1969) *The Economy of Cities* (New York: Vintage).
Johnson, S. (2010) *Where Good Ideas Come From* (New York: Riverhead).
Jørgensen, U. & Karnøe, P. (1995) The Danish wind turbine story: Technical solutions to political visions, in: A. Rip, T. Misa & J. Schot (Eds) *Managing Technology in Society: The Approach of Constructive Technology Management* (London: Pinter).
Karnøe, P. & Jorgensen, U. (1996) *The International Position & Development of the Danish Wind Turbine Industry* (Copenhagen: AKF).
Kauffman, S. (1995) *At Home in the Universe* (Oxford: Oxford University Press).
Kauffman, S. (2008) *Reinventing the Sacred* (New York: Basic Books).
Kemp, R. (2002) Environmental protection through technological regime shifts, in: A. Jamison & H. Rohracher (Eds) *Technology Studies and Sustainable Development* (Munich: Profil Verlag).
Klepper, S. (2002) Capabilities of new firms and the evolution of the US automobile industry, *Industrial & Corporate Change*, 11, pp. 645–666.
Laclau, E. & Mouffe, C. (1985) *Hegemony & Socialist Strategy: Towards a Radical Democratic Politics* (London: Verso).
Lundvall, B. (1988) Innovation as an interactive process: From user–producer interaction to the national system of innovation, in: G. Dosi, C. Freeman, R. Nelson, G. Silverberg & L. Soete (Eds) *Technical Change & Economic Theory* (London: Pinter).
Martin, R. L. (2010) The Roepke lecture in economic geography—rethinking regional path dependence: Beyond lock-in to evolution, *Economic Geography*, 86, pp. 1–27.
Martin, R. L. (2011) Regional economies as path dependent systems: Some issues and implications, in: P. Cooke, B. Asheim, R. Boschma, R. Martin, D. Schwartz & F. Tödtling (Eds) *The Handbook of Regional Innovation & Growth* (Cheltenham: Edward Elgar).
Martin, R. L. & Sunley, P. (2006) Path dependence and regional economic evolution, *Journal of Economic Geography*, 6, pp. 395–438.
Martin, R. L. & Sunley, P. (2010) The place of path dependence in an evolutionary perspective on the economic landscape, in: R. Boschma & R. L. Martin (Eds) *Handbook of Evolutionary Economic Geography*, pp. 62–92 (Cheltenham: Edward Elgar).
Ministry of Foreign Affairs of Denmark (2008) *Invest in Denmark* (Copenhagen: Government of Denmark).
Neffke, F. & Svensson Henning, M. (2008) Revealed relatedness: mapping industry space, Papers in Evolutionary Economic Geography 0819, University of Utrecht, Section of Economic Geography.
Nelson, R. (1994) The co-evolution of technology, industrial structure, and supporting institutions, *Industrial and Corporate Change*, 3, pp. 47–63.
Nelson, R. (2008) What enables rapid economic progress: What are the needed institutions? *Research Policy*, 37, pp. 1–11.
Page, S. (2006) Path dependence, *Quarterly Journal of Political Science*, 1, pp. 87–115.
Pisano, G. & Verganti, R. (2008) Which kind of collaboration is right for you? *Harvard Business Review*, 86, pp. 78–86.
Schumpeter, J. (1975) *Capitalism, Socialism & Democracy* (New York: Harper Torchbooks).
Storper, M. & Walker, R. (1989) *The Capitalist Imperative: Territory, Technology & Industrial Growth* (Oxford: Blackwell).
Sydow, J., Schreyogg, G. & Koch, J. (2009) Organizational path dependence: Opening the black box, *Academy of Management Review*, 34, pp. 689–709.
Verganti, R. (2006) Innovating through design, *Harvard Business Review*, December, Reprint R0612G, pp. 1–9.
Verganti, R. (2009) *Design Driven Innovation* (Boston: Harvard Business Press).
Von Hippel, E. (1988) *The Sources of Innovation* (Oxford: Oxford University Press).
Von Hippel, E. (2005) *Democratizing Innovation* (Cambridge: MIT Press).
Wenting, R. (2009) *The Evolution of a Creative Industry* (Utrecht: University of Utrecht Press).

From the Old Path of Shipbuilding onto the New Path of Offshore Wind Energy? The Case of Northern Germany

DIRK FORNAHL*, ROBERT HASSINK**, CLAUDIA KLAERDING**, IVO MOSSIG† & HEIKE SCHRÖDER†

*Centre for Regional and Innovation Economics at the University of Bremen, Germany, **Department of Geography at the University of Kiel, Germany, †Department of Geography at the University of Bremen, Germany

Abstract *Wind energy-related employment has been surging recently in Germany: it rose from 9200 in 1997 to 90,000 in 2007 and is estimated to be 112,000 in 2020. The industry particularly emerged in coastal, Northern Germany. Recently, big hopes have been particularly set on the offshore wind energy industry. Two recently discussed evolutionary concepts explain the emergence of new industries, such as wind energy, in space: the windows of locational opportunity concept stresses the locational freedom in the earliest stages of industrial development, whereas path creation emphasizes the role of existing industrial development paths, such as shipbuilding, from which new industrial paths, such as wind energy, emerge. This paper aims at analysing whether the new industrial path of offshore wind energy emerged out of existing paths, mainly shipbuilding, in the five states of coastal Germany, namely Bremen, Lower Saxony, Hamburg, Schleswig-Holstein and Mecklenburg-Vorpommern. It concludes that shipbuilding only indirectly affected the emergence of the new industrial development path of the offshore wind energy industry in Northern Germany.*

1. Introduction

After the Second World War, the shipbuilding industry became the core manufacturing industry in Northern Germany.[1] Several yard crises starting in the 1980s led to pressures to restructure the regional economy. Against this background, big hopes have been set on the booming offshore wind energy sector as a new impetus for regional employment and economic growth. Schamp (2000, p. 136), though, has pointed out that technologically determined industry cycles are not the only explanation for the decline and renewal in old industrial areas. Not only industries' age, but also institutional tissues, sclerotic networks, institutional persistence and other lock-in effects have to be broken up in order to use the potential of new technologies and to master the challenges of industrial restructuring.

The success and regional embeddedness of a new industry such as the offshore wind energy are affected by a combination of factors (Martin & Sunley, 2010; Boschma & Frenken, 2011; Storper, 2011). In addition to the motivation of decisive actors and market opportunities, they are particularly affected by specific capabilities which can differ from region to region due to varying regional conditions. This paper particularly focuses on the role of these regionally specific conditions related to the previously dominating shipbuilding industry in explaining the genesis of the offshore wind energy industry in the five states of Northern Germany (Lower Saxony, Bremen, Hamburg, Schleswig-Holstein and Mecklenburg-Vorpommern) (see Figure 1). It extends both the empirical and theoretical parts of a previously published paper in German which was confined to Lower Saxony and Bremen (see Mossig *et al.*, 2010). The questions that we would like to tackle are as follows: do traditional shipbuilding companies invest in the new business fields of offshore wind energy? Which expertise, competences, skills and routines of former yard workers and engineers can be recombined and used in offshore wind energy? Are traditional network relations extended in the new field of offshore wind energy? At which points in the industrial development path and how did regional industrial policy in the five regional states (Länder) intervene in the restructuring process? Does the existing infrastructure constitute an important locational advantage in supporting the emergence of the offshore wind energy industry in the different states of Northern Germany? Or are exogenous factors more important, such as the attraction of energy companies or key actors from outside the region, to explain the emergence of the industry in the regions? Overall, to what extent is the emerging industrial development path of offshore wind energy in Northern Germany related to the decline of the shipbuilding industry? Can we speak of tight relations with many commonalities or can we explain the emergence of the industry better by new impetuses and new windows of opportunity?

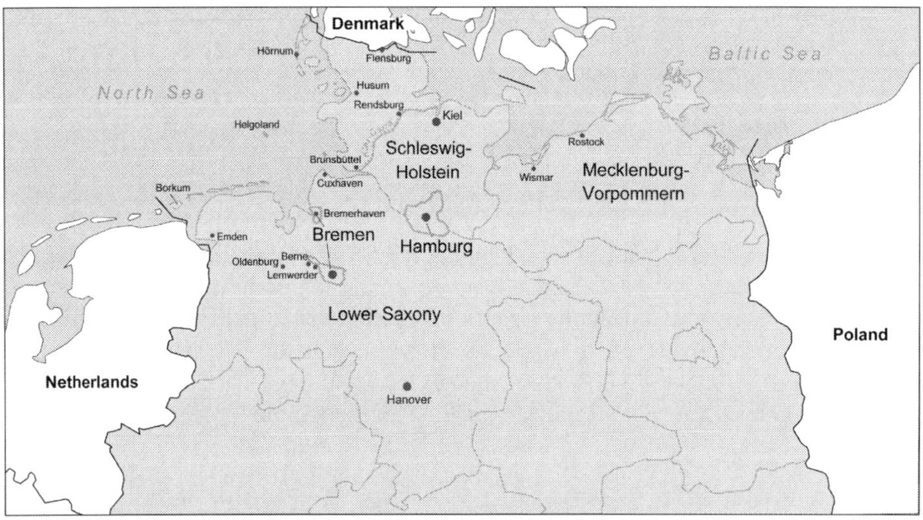

Figure 1. Map of Northern Germany.

The perception of a smooth restructuring from a declining core industry into a newly emerging growth industry, however, can also be the result of wishful thinking of regional economic promoters or local politicians. A handful newspaper articles about individual shipbuilding companies which were saved from bankruptcy due to offshore wind energy orders or easily made correlations between yards and offshore terminals might hide that shipbuilding-related region-specific conditions are in fact not very relevant to offshore wind energy. Perhaps other triggering events have been much more significant. These might include the establishment or relocation of a core company, newly developed product innovations, specific policy measures as a reaction to the yard crises or the activities of new actors in the regional institutional environment and their effective promotion of offshore wind energy (Brenner & Fornahl, 2006). Northern Germany can be seen as an appropriate research area to tackle these questions, as in this region, we can find both the declining shipbuilding industry and a relatively high density of research institutes and companies in offshore wind energy (see, for instance, IWR, 2008). The main sources for the empirical parts of this research (Sections 2 and 4) consist, on the one hand, of primary data collected through 17 in-depth interviews with the main actors in the states of Northern Germany, such as the leading offshore wind energy companies, suppliers, shipbuilding companies, cluster managers, officials of industry associations and officials in state ministries (for reasons of confidentiality, we cannot disclose the names of the interviewees). On the other hand, we have also used secondary data in the form of annual reports, strategy papers and other statistics and reports. The interviews were carried out by the authors of this paper. The first six interviews in Lower Saxony and Bremen took place in March 2010, whereas in August 2010, another 11 followed in Schleswig-Holstein, Hamburg and Mecklenburg-Vorpommern. The interviewees were selected with the help of cluster managers in the different states.

The paper is structured as follows. In the next section, the decline in shipbuilding and emergence of the offshore wind energy industry is further described. In Section 3, the theoretical foundations of this paper based on path dependence and path creation are introduced, which have recently been much discussed in economic geography (see, for instance, Martin & Sunley, 2006; Martin, 2010). In Section 4, the empirical analysis of the relationship between the emerging offshore wind energy industry and the declining shipbuilding industry in the different states of Northern Germany is presented. In Section 5, summary and conclusions are given.

2. Offshore Wind Energy as the Great White Hope after the Yard Crises

In the 1950s, the shipbuilding industry in the northern part of former West Germany went through a real boom. Despite heavy war-related damages, the industry boomed mainly due to tax reductions and favourable loans in order to reconstruct the trade fleet. The worldwide intensification of trade relations generated strong demand for large trade merchant vessels and hence the employment increased up to 113,000 in 1958 (200,000 if one would include suppliers); the lion's share was employed in the coastal areas of West Germany. West Germany's world market share in merchant shipbuilding reached its peak in 1956 with 17%. At that time, it was the third biggest shipbuilding nation in the world, after Japan and Great Britain (Nuhn, 1990; Giese *et al.*, 2011).

In the 1960s, business cycles started to show first ups and downs, as well as first bankruptcies and mergers and acquisitions of small- and medium-sized yards. A dramatic decline started at the end of the 1970s. In 1978, West Germany's world market share in

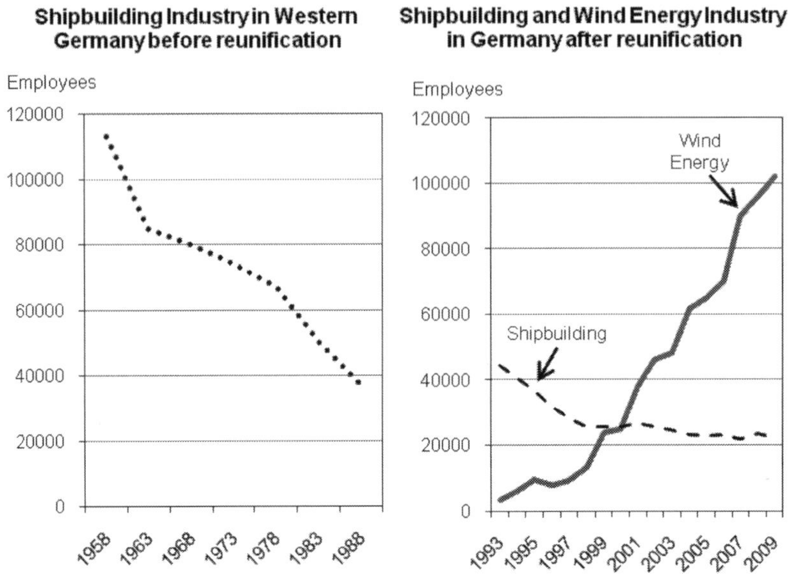

Figure 2. Development of employment in the shipbuilding industry in Western Germany before reunification (1958–1988) as well as in the shipbuilding and wind energy industries in Germany (1993–2009). *Source*: Nuhn (1998, p. 321) as well as annual reports of the German Shipbuilding and Ocean Industries Association and the BWE.

merchant shipbuilding had dropped to only 2.1%. Despite strong support from the federal and state governments, only 59% of the production capacities were used in 1979 (Nuhn, 1990). Although East Germany's shipbuilding located in the state of Mecklenburg-Vorpommern was added to total employment in shipbuilding after reunification in 1990, the dramatic decline in yard employment steadily continued to the mere 22,000 employment currently (VSM, 2010, p. 65) (Figure 2). Large traditional shipyards, such as Bremer Vulkan, which employed over 5000 workers in the 1970s, as well as many small- and medium-sized yards, had to give up in the 1990s. The main reasons for this strong decline in the shipbuilding industry were seen in the strongly developing international competition in East Asia, the delay in necessary restructuring of the product range partly caused by state subsidies and relatively high production costs due to high wages and material costs, as well as currency disadvantages (strong D-Mark) (see Kramm, 1980; Nuhn, 1998, p. 318; Eich-Born, 2005; Eich-Born & Hassink, 2005; Hassink & Shin, 2005; Tholen & Ludwig, 2005; Giese *et al.*, 2011). Of the currently employed 22,000 people in German shipyards, almost a third (6724) are located in Lower Saxony, followed by those in Mecklenburg-Vorpommern (4916), Schleswig-Holstein (4565), Hamburg (2565) and Bremen (1443) (VSM, 2010).

Recently, high hopes have been set on the booming offshore wind energy industry in Northern Germany. According to the German Wind Energy Association (BWE), only 9200 persons were employed in the planning and construction of wind farms in 1997, whereas about 10 years later, the number increased to approximately 90,000 (Figure 2) (for a general overview on the wind energy industry, see Musgrove, 2010). Until 2020, the total number of employees is expected to be 112,000. These positive future expectations are mainly due to the strong expansion of the number of offshore wind farms. After finishing the construction of the first offshore wind farm, Alpha Ventus, 45 km north of Borkum in the North Sea, in November 2009, further investments in infrastructure are expected in three areas.

First, together with neighbouring countries, billions of euros will be invested in the grid network on the bottom of the North Sea.[2] Second, investments will be made in the extension of existing harbours with special terminals for shipping the extremely heavy and huge components of offshore wind farms. In immediate vicinity to these harbours, industrial estates are planned to provide wind energy companies with manufacturing premises. The investments for the new offshore harbour of Bremerhaven, to ship annually between 110 and 160 installations, are, for instance, estimated to be between €170 and €200 million (Weser Kurier, 2010b). About €80 million will be invested in the harbour of Cuxhaven (Lower Saxony), whereas other investments are planned to extend the harbours of Emden (Lower Saxony), Brunsbüttel, Husum and the county-port of Rendsburg (all in Schleswig-Holstein). Originally a port for reloading, Rendsburg now invests in extending industrial estates close by to become a heavy-duty harbour in 2011. The area called Neuer Hafen Kiel-Canal is advertised for heavy-weight productions of traditional machine-building and explicitly geared to producing and assembling components of the wind energy industry. Moreover recently, increasing co-operation between the relatively small ports in Schleswig-Holstein has been observed, in order to come to a functional division of labour concerning offshore wind energy transport and repairing (Kieler Nachrichten, 2011b). In Mecklenburg-Vorpommern, there have also been heavy investments in the port infrastructure of Rostock since 2003, albeit with a focus on port-related activities in general, not particularly on the offshore wind energy sector.

Third, private money as well as public grants will be invested in the extension of the research infrastructure. In Lower Saxony and Bremen, for instance, the universities of Bremen, Hanover and Oldenburg, as well as the Fraunhofer Institute for Wind Energy and Energy System (IWES), have joined forces under the name ForWind in order to cover all wind energy-related research in the north-western part of Germany. In Schleswig-Holstein, a similar initiative was set up under the name CEwind. In Bremerhaven, in particular, the networking between firms has been boosted by the establishment of the Wind Energy Agency Bremerhaven/Bremen (WAB). In Husum, Schleswig-Holstein, it is Windcomm, and in Mecklenburg-Vorpommern, there is a similar organization called Wind Energy Network. An additional network organization is Germanwind—the wind energy cluster in northwest Germany, which has presently 150 companies and institutes as members. All in all, offshore wind energy already is a strong regional pillar in large parts of Northern Germany, particularly in the areas around Bremen (Seebeck, 2011), on which large hopes are set to trigger strong economic growth in the future and to compensate for the huge job losses related to the yard crises, although there is a considerable time lag between the decline of the shipbuilding industry and the emergence of the offshore wind energy industry (Zeit, 2010). The question is, however, how did it emerge? And can its origin be explained by theories on the origin of new development paths?

3. Conceptual Framework: Explaining the Origin of New Industrial Development Paths

Path dependence-related ideas have already been developed in Carl Menger's analysis of institutional emergence in 1883 as well as in Thorstein Veblen's concept of cumulative causation in the evolution of habits and conventions (1898). It was only in the 1980s, however, that path dependence received much more attention, particularly due to the work by Paul David and Brian Arthur (Martin & Sunley, 2006, p. 397). In their work, they opposed neo-classical assumptions and explained multiple equilibria and market inefficiencies with the help of path dependence (for instance, Arthur, 1988, 1994; David, 1993; Crouch & Farrell, 2004, pp. 8–10).

A path-dependent process is characterized by non-ergodicity. This term, which has been adopted from the theory of stochastic systems, means that actors or systems which are subject to path dependence cannot free themselves from the effects of past events (David, 1993, p. 29; Martin & Sunley, 2006, p. 399). Path dependence is not the same as inertia or persistence effects, it refers instead to the limitation of options for current decisions, as they are strongly related to events and experiences made in the past (North, 2005, p. 52). At the same time, current decisions are dependent on the current and contingent context, so that the concept of path dependence can be clearly distinguished from historical determinism (Bathelt & Glückler, 2002, p. 27 ff).

A perspective of path dependence can contribute to understanding and analysing the origin of the offshore wind energy industry in Northern Germany in the context of the region's industrial restructuring problems. In order to do so, the following two questions need to be first tackled at the theoretical level:

(a) As far as there are only limited relations with the shipbuilding industry, the question about the origin of a new industrial development path needs to be dealt with, which will be done in Section 3.1.
(b) If there are a large variety of relations, it needs to be clarified how the development path of the traditional core industry can be transferred and how inertia which caused the decline of the shipbuilding industry can be de-locked (see Section 3.2).

3.1 *The Origin of New Industrial Development Paths*

In the literature, often existing industrial development paths are investigated, whereas the question of the actual origin of a new industrial path is not much dealt with (Hirsch & Gillespie, 2001, pp. 72, 84; Crouch & Farrell, 2004, p. 7; MacKinnon *et al.*, 2009, p. 143). Basically, three different viewpoints can be distinguished on the origin of new industrial development paths. The causes of a new path are first seen in chance, second, in a mix of chance and limiting conditions and, third, in the deliberate and dedicated behaviour of economic actors. The first viewpoint, that the origin of new development paths is a matter of pure chance, does not need a further theoretical elaboration. The second viewpoint is often backed by the theory of windows of locational opportunity, which is used in evolutionary economic geography (Martin & Sunley, 2006). According to this theory, innovative enterprises of new industries do not have specific locational preferences in the early stages of development yet. They shape their entrepreneurial environment largely themselves and therefore have large degrees of freedom concerning their choice of location. In these early stages of development, therefore, the windows of locational opportunity are relatively open and hence existing locational patterns can be changed. Chance and small events determine where the first entrepreneurs in a new industry will locate. Also, regions outside of existing economic centres get a chance to generate or attract pioneer entrepreneurs as long as certain unspecified framework conditions and basic requirements are in place (Storper & Walker, 1989, pp. 70–92; Boschma, 1997, p. 15 f.; Bathelt & Glückler, 2003; Mossig, 2006, p. 58ff.; Dorenkamp & Mossig, 2010). According to Boschma and Frenken (2006, p. 290), "... regional conditions may play a generic and rather unimportant role at the start of a new sector, such as providing generic knowledge and skills, functions that are often equally well provided in many other regions". The third viewpoint criticizes the strong emphasis on chance and the lacking integration of actor and behavioural theories in the concept of path dependence (Garud & Karnøe, 2001; Garud *et al.*, 2010). According to them, new development paths are created by strategic actions of actors who deliberately differ themselves from existing social rules and technologies. According to this viewpoint, the location of pioneer entrepreneurs as seed-beds of a new development cannot be explained by chance and small events, but is strongly related to existing experience, knowledge, capabilities and contacts, which the entrepreneur carries with him or her from previous industrial development paths.

The question, therefore, is when do we speak of a truly new path and when of a change of an existing path? According to Karnøe and Garud (1995, cited in Rao & Singh, 2001, p. 243 f.), a new path is created by a sudden break with the past and hence a sudden break with existing technologies, products and organizational forms, which implies new future expectations. Although we realize that path creation refers to more than new industries alone, our focus is on path dependence and path creation in relation to industries, to which we refer to as industrial development paths. In our view, there are strong inter-relationships between breaks with existing technologies, products and organizational forms and the formation of new industries, but not necessarily so, as in the same industry, there may be a technological break or change, or the other way around, the same technology may play a role in different industries (we are grateful to one of the reviewers for making this point).

Karnøe and Garud (1995) do not speak of an incremental adaptation process over a long period of time in which the different elements of a path are gradually changed by the actions and experiments of the actors, which could be characterized as an on-path change. Accordingly, it needs to be clarified which triggering events lead to such a path-breaking change. In contrast to the above-mentioned small events (see also Hirsch & Gillespie, 2001, p. 72; Jovanović, 2009, p. 71), we need a much stronger driving force in order to break through an established path (David, 1993, p. 39). Technological breakthroughs are often mentioned as such driving forces in the economic geography literature, in line with the long wave theory and neo-Schumpeterian approaches (Boschma, 1997, p. 13; Boschma & Lambooy, 1999, p. 421) and the model of industrial pathways of Storper and Walker (1989). We, therefore, need to check whether the offshore wind energy can be regarded as such a technological breakthrough or as an incremental change.

Martin and Sunley (2010) summarize the discussion on the different opinions on the origin of new industrial path development with four quadrants in a matrix (see Table 1). They state (Martin & Sunley, 2010, pp. 79, 80) that

> while the classic model of path dependence is rooted in quadrant 4, more recent work in economic geography has begun to move closer to positions in quadrants 2 and 1, and has put much more emphasis on the re-use and transfer of resources and competences.

With regard to the links between old and new paths, Martin (2010, p. 19) states that

> . . . new paths may be latent in old ones, or spin out from existing ones . . . resources and competences acquired and used in previous and existing paths of technological and industrial activity may be recombined to form the basis of purposeful entrepreneurial deviations into new paths . . . preconditions, and the resources associated with them, are often place specific, shaped by the characteristics of previous local economic developments . . . the local inherited knowledge and skill base of an industry can form the basis of the rise of related new local paths of industrial and technological activity. And local spin-off firms can use the routines and competences inherited from their parent firms to launch new products and processes.

Bathelt and Boggs (2003, p. 256 f.), however, emphasize that it is not sufficient to just focus on technological discontinuities as an explanation for the destruction or emergence of development paths. They stress that externally caused strong crises also have the potential to break through the path dependence of economic development (Bassanini & Dosi, 2001, p. 50; Jovanović, 2009, p. 13; MacKinnon *et al.*, 2009, p. 143). Moreover, specific political support measures potentially have the power to break through existing paths. Particularly those policy-makers who are not themselves part of the old development path (e.g. the local administration) are able to decrease incentives to follow the old path or increase incentives to change towards a new development path. This might, for example, be achieved by legislations positively or negatively affecting the market conditions under which the different regimes are operating.

In addition to the question whether an old path needs to be broken through to create a new one, we also need to think about factors affecting the development process of a new path in a positive way. Four factors can be mentioned (see also Brenner & Fornahl, 2006, in the context of clusters). First, the current market conditions and forecasts need to be sufficiently positive, so that entrepreneurs are willing to enter this market. Second, long-term stable characteristics of a region (such as its coastal location) do play a role. Third, varying local conditions such as human capital, networks and public research establishments also play a role. Fourth, there need to be triggering events leading to an actual realization of the regional and industrial potentials, such as the location of a core company or new industry policy initiatives.

Building upon the current literature on the origin of new development paths, we now need to examine whether the yard crisis in Northern Germany can be regarded as a path-breaking event and whether this event has been the core impetus for the new development path of the offshore wind energy industry. We analyse which combination of deliberate actions, region-specific conditions and triggering events as well as geographical conditions has played a role here. Alternatively chance and small events might play a role as is postulated in the theory of windows of locational opportunity. They need to be interpreted as independent of the traditional shipbuilding development path and as behaviour that is not deliberately focused on establishing an offshore wind energy cluster.

After having dealt with the origin of new development paths from a theoretical perspective, we now elaborate more on the mechanisms leading to a breakthrough of existing development paths and a successful further development of the newly created path.

3.2 *De-locking of Industrial Development Paths*

In order to realize a link between the two development paths (shipbuilding and offshore wind energy), the new industry needs to be able to free itself from the declining core indus-

Table 1. Varieties of path creation.

	Origins of a new path of development	
Place and path effects	Deliberate and intentional	Chance and accidental
Enabling new paths	1. Agents search for opportunities, re-use resources and transfer competences as the basis of new growth	2. Agents gain assets and experience, but accidents and events trigger a new path
Constraining to existing path	3. Designed interventions to break path or switch location to overcome lock-in	4. Unpredictable external shocks and random events break old trajectory and launch a new path

Source: Martin and Sunley (2010), p. 80.

try. Sclerosis is regarded as an important reason for not successfully coping with the yard crisis in Northern Germany and many other old industrial areas for that matter (Hassink, 2010). State subsidies for the yards in West Germany, for instance, led to conserving traditional products and hence a delaying of necessary restructuring into new, high-value-added special products. State subsidies were used to "winter" crises until the next boom. This strategy led to sclerosis of production and network structures leading to a negative lock-in (Kern & Schumann, 1990, p. 305ff.).

A large part of the literature on lock-ins discusses the factors leading to a negative locking in of development paths and thinks about strategies on how to prevent this (for instance, Grabher, 1993; Cowan & Gunby, 1996; Boschma, 2004; Hassink, 2005, 2010). Relatively little has been written on how regions can manage to free themselves from such a lock-in situation. Concerning de-locking, most economic geographers have until now emphasized the technological level of this process. In addition to technologies, heterogeneity, on the one hand, and transplantation through the import of novelties from outside the action radius of the company, on the other hand, are useful to be discussed in relation to de-locking (Martin & Sunley, 2006, pp. 121–123).

Heterogeneity as a first de-locking mechanism is a relatively new research area. In evolutionary economics, heterogeneity is mainly discussed at the micro-level of individual firms as a driving force in selection processes (Rigby & Essletzbichler, 1997, p. 270). Frenken *et al.* (2007) have recently stressed the meaning of heterogeneity of different mutually related industries in a region with the term related variety. They particularly emphasize positive effects of knowledge spillovers, risk reduction and labour market stability. Heterogeneity, however, does not seem to be very relevant for our research question, that is, the relation between the two development paths, shipbuilding and offshore wind energy industries.

The decisive impulses for the emergence of offshore wind energy industry might have also come from actors external to the industry. This leads us to the second de-locking mechanism, namely transplantation. Knowledge acquisition, learning, innovativeness and hence competitiveness of firms nowadays do not only depend on intra-organizational resources, but increasingly also depend on extra-organizational resources (van den Bosch *et al.*, 1999, p. 552; Song *et al.*, 2003, p. 351 f; Boschma, 2004, p. 1004; Martin & Sunley, 2006, p. 422; Bierly *et al.*, 2009; p. 481 f.). In order to break through sclerotic development paths, it does not suffice to recognize external relevant knowledge, what is also needed is the capability in firms to synthesize newly acquired knowledge with existing knowledge stocks, which is in turn dependent on the so-called absorptive capacities in firms (Zahra & George, 2002, p. 185; Phene & Almeida, 2008, p. 902). Transplantation is seen as an important de-locking mechanism due to the learning processes that take place through the exchange with actors from other institutional contexts. Learning processes as such, however, have controversial effects, as they are both path dependent and path breaking at the same time (Crouch & Farrell, 2004, p. 26 f., p. 34 f.; Martin & Sunley, 2006, p. 422 f.). In the empirical part of this paper, we will clarify the contribution of transplantation of external knowledge to the breakthrough of the development path of the shipbuilding industry and the emergence of the offshore wind energy cluster. In this case, probably not only know-how transfer, but also the transfer of external capital has been relevant.

4. Determinants of the Genesis of the Offshore Wind Energy Industry in Northern Germany

In the following sections, the origin and development path of the offshore wind energy cluster in Northern Germany will be described and interpreted with the help of the above-described theoretical elaborations. They can be regarded as a thick, qualitative description of the genesis of the offshore wind energy industry in Northern Germany based on a combination of interview results and secondary data (for further details, see Section 1).

4.1 *Local Conditions Affecting the Offshore Wind Energy Industry*

Two shipbuilding-related local factors which have positively affected the location of offshore wind energy companies in Northern Germany can be distinguished.

First, the available infrastructure (e.g. large dedicated quay premises, production halls, harbour facilities and heavy lift terminals) has been a favourable location factor for offshore wind energy companies. Large premises and production halls are needed for the production, storage and pre-assemblage of offshore components, such as 60-m-long rotor blades. Some large quaysides of bankrupt shipyards were made available for these purposes, although small adaptations to the specific requirements of the offshore wind energy companies were necessary. AMBAU Stahl- und Anlagenbau, for instance, is located on premises of the former Vulkan yard in Bremen Nord and WeserWind GmbH Offshore Construction Georgsmarienhütte at a former yard in Bremerhaven.

Second, the locally available human capital played a positive role. Regional competences in firms and workers are available in steel construction (for instance, welding and assembling of large components), electronics in autonomous systems, maritime logistics and the handling of heavy-weight components. Most yards used to have relatively small shares of in-house production, which has led to an increase in the necessary competences in co-ordination and logistics, competences which can also be used in the offshore wind energy industry. Due to the yard crises, parts of these human resources have been made redundant and could be used in the offshore sector. WeserWind GmbH Offshore Construction Georgsmarienhütte, for instance, has employed workers from SSW Transfergesellschaft, a job intermediary agency founded after the bankruptcy of Werft SSW Schichau Seebeck Shipyard. SIAG Schaaf Industrie AG has taken over a large number of the yard workers of TKMS Nordseewerke in Emden. Other shipyards in Schleswig-Holstein and Mecklenburg-Vorpommern, however, substantially rely or plan to rely on existing competences and specialized in-house training for building installation ships and transformation platforms required for the wind energy sector. At the same time, however, the pool of workers is limited as some yard workers moved to other regions or have retired before the offshore wind energy industry started to emerge. During the last few years, the number of retraining and additional training courses for offshore activities has increased, in both the craft and academic fields. In Schleswig-Holstein, some public and private institutions such as Glücksburg Consulting Group are engaged in retraining personnel from related sectors of the onshore and maritime industry. However, there are also further potentials for improving the training infrastructure in Schleswig-Holstein, particularly for offshore service technicians. Past training courses on safety were, for instance, offered in Esbjerg in Denmark. Although many former yard workers have found a job in the offshore wind industry, interviewees stressed that the capabilities and competences could also have come from an industry other than the shipbuilding industry. A basic, high-quality craft education is sufficient for employees to be retrained into specific qualification requirements of the offshore wind industry. Although regions with yards have some favourable local conditions, similar conditions could have been found in other regions with different industrial structures. Human capital as such, therefore, cannot be seen as sufficient evidence for new path creation out of established paths.

The direct access to open sea due to the availability of deep-water seaports has been an essential factor contributing to the emergence of both the shipbuilding and offshore wind energy industries in Northern Germany. Transportation costs are much higher in alternative locations without a direct access to open sea, particularly due to heavy-weight components, such as foundations and piles. In order to save transportation costs, producers of both offshore and export-oriented onshore components, therefore, prefer waterfront locations along the sea or large rivers, such as the Rhine and Elbe. Supply, repair and maintenance of the wind farms in the North Sea and Baltic Sea will also be co-ordinated and started in the offshore wind energy seaports. All in all, sea access as a local condition is favourable to both industries; the shipbuilding industry did positively but rather indirectly affect the emergence of the offshore wind energy industry because of this.

In addition to the favourable local conditions which have been partly created by the shipbuilding industry in the past (for instance, waterfront premises and heavy lift facilities), policy-makers and companies have more recently taken dedicated actions to improve local conditions even more. Concerning actions taken by policy-makers, the city state of Bremen has been the first in Northern Germany and it started its offshore wind energy support policy in 2001. This policy consists of R&D and investment support schemes, as well as support for networks and offshore-oriented infrastructure. It made, for instance, dedicated premises available for offshore activities and invested in the adaptation of seaport infrastructure to offshore activities. Concerning network activities, informal meetings started in 2001, first primarily with research institutes, later on with support for lobby activities and public relations. This led to the establishment of the WAB in 2004. Since the offshore wind energy industry is still relatively small and regional policy-makers in this area started to support the industry relatively early, tight personal networks could be built up between policy-makers, public administrators and company managers. Due to the early policy involvement, as well as the broad political support from several administrative departments, a visible, positive sign was given to potential investors in the Bremen region. A strong dedication to this industry by regional policy-makers in Bremen was conveyed to companies willing to relocate their offshore activities or to companies planning to set up an offshore daughter. This political dedication and the favourable local conditions actually led to the location and relocation of some offshore wind energy companies to the Bremen region that came from outside the region. Company interviewees positively report about the leeway they have in the Bremen region to change local conditions into favourable ones for the offshore wind energy industry. Schleswig-Holstein and Mecklenburg-Vorpommern, in contrast, have reacted much later and could hence not be as successful as the Bremen region in attracting offshore wind energy companies from outside. Only since 2009, the wind energy sector has become a prioritized field of economic development in Schleswig-Holstein (Kieler Nachrichten, 2010; Windcomm, 2010). For offshore wind energy, there is no specific support programme yet, but individual initiatives, such as CEwind, are supported from the overall Future Programme Economy.

The policy support measures, however, have been carried out separately from the yard crisis. Although policy-makers have been looking for alternative industries to compensate for the job losses in shipbuilding already since the 1990s in all northern states (push factor), the heaviest job losses occurred decades ago (in West Germany), much earlier than the emergence of the offshore wind energy industry. Moreover, offshore is not the only alternative industry selected by policy-makers; there is also support for a couple of other industries, such as the tourism industry and the creative industry (Haller *et al.*, 2003; Dirksmeier, 2009).

Also, offshore wind-related public research activities are largely separated from the traditional development path of shipbuilding. They mainly emerged out of existing research institutes outside of shipbuilding-related research, such as the Fraunhofer Institute for IWES in Bremerhaven and the Fraunhofer Institute for Manufacturing Technology and Advanced Materials in Bremen, or research activities in (onshore) wind energy companies. Initially, these research activities were not focused on offshore or did take place in an unco-ordinated way. In order to improve this situation and to bundle offshore research activities, the research and co-ordination agency at the Hochschule Bremerhaven (fk-wind) was established in 2003. Moreover, new institutes that particularly focus on offshore-related research were set up in Bremen. According to a study on renewable energy in Germany, both the Rhine-Ruhr-Region and Bremen/Bremerhaven are the main centres of wind energy research in Germany (IWR, 2008). In addition to research, bachelor and master studies on wind energy technology have also been established at the Bremerhaven University of Applied Sciences. In Schleswig-Holstein, similar initiatives have been started since 2005, such as CEwind and the master programme of wind engineering. Launched by the University of Applied Science of Kiel and the University of Applied Science of Flensburg in 2008/2009, a tailor-made course targeting emerging market demands has been designed. In Mecklenburg-Vorpommern, similar master programmes will run from 2011 onwards at the University of Rostock.

Another indicator for the emergence of a new path separated from the traditional development path of shipbuilding is the origin of some offshore wind energy companies. Some companies, for instance, developed out of the onshore wind energy industry and plastic processing and the offshore oil and gas industries. Investments in new competences are necessary, also in onshore wind energy companies, as offshore has its specific requirements concerning corrosion protection, maintenance intervals and specific services, as well as specific installation and maintenance logistics. Moreover, some equipment such as pile-driving hammers and pipes need to be customized to specific demands and often cannot be re-utilized for other projects or be produced in series. Offshore project solutions, therefore, have to be individually designed in line with specific industrial requirements and expertise.

The financial sector has been sceptical vis-à-vis investments in offshore wind farms, particularly at the early stages of development; in contrast to those of onshore wind farms, both investment sums and equity ratio are higher and the industry first had to prove itself. Also, as procedures of building transformation platforms, hammers and installation ships have not been standardized yet, the calculation of risks is claimed to be too vague for supporting huge investment decisions. Until the yard crisis, investing in shipbuilding was a safer bet for banks. As a reaction to this sceptical attitude of the banks, experienced onshore product developers (e.g. wpd, PROKON or Energiekontor) saved financing the first project by selling loans, a strategy also used in the shipbuilding sector to finance large projects. In the wind energy sector, shipyards, therefore, appear to be integrated into the value chain as providers of capital rather than as technical constructors and producers of installation ships. More recently, however, large energy concerns, such as Eon, EnBW and RWE, and investment funds as well as banks increasingly started to invest in offshore projects, as it got easier to calculate returns on investments due to the Renewable Energy Sources Act (EEG) and to forecast wind yields.

As with banks, the industry had to prove itself vis-à-vis potential suppliers in order to be able to meet the demand for specific supplies and components in the region. In some cases, networks and competences that emerged out of the shipbuilding industry could be used.

4.2 *Market Incentives as Driving Force of the Industrial Development Path*

The growing market has played a key role in developing the offshore wind energy industry. Important factors have been the grid connection duty, the EEG in 2004, an increasing environment consciousness, the Infrastructure Planning Acceleration Act, oil price increases, climate protection agreements and increasing doubts vis-à-vis nuclear energy, as well as the high-tech strategy of the federal government. The latter has proclaimed the area of climate protection, resource protection and energy as the leading future market in 2006. This has led to, first, a stronger support for R&D in wind energy and positive effects on the supply side. Second, this has also positively influenced the demand side. Moreover, increasing feasibility and risk calculation of offshore wind farm projects have been made possible due to the long-term stability of the feed-in tariffs at 15 cents/kWh for 12–20 years. These positive framework conditions could hardly be affected by industrial and/or regional actors, apart from some lobbying activities, for instance, to influence the EEG Act. Nevertheless, the industry did not emerge by chance, but its emergence is the logical consequence of a series of events, starting already with the oil crises in the 1970s, and deliberate actions to change societal, ecological, economic and political framework conditions. Despite these positive framework conditions, there have been several uncertainties. It took much longer to open the first test farm, Alpha Ventus, which had been approved already in 2001 but started only in 2009. The interviewees confirmed that it is particularly the long-term stability of the feed-in tariffs at 15 cents/kWh, which was agreed upon in 2009, that significantly reduced economic uncertainty.

4.3 *The Origin of Leading Offshore Wind Energy Companies*

It is not only in times of crises that companies, such as shipbuilding companies, look for new business areas. The motivation behind this searching behaviour can be seen in the positive and negative effects of displacement (Mossig, 2000). In order to cope with the yard crisis, offshore wind energy was not in the focus of the shipbuilding companies in their search for alternative business areas. Other areas were tested to find out whether existing competences could be transferred in a profitable way. Only a few exceptional shipyards considered offshore wind energy as an alternative business area and hence contributed to the endogenous emergence of offshore wind energy in the region: Husumer Schiffswerft, FR. Fassmer (Berne) and Abeking & Rasmussen in Lemwerder. In some cases, such as Abeking & Rasmussen, there were competences available in the area of fibre-reinforced composites at the end of the 1980s, for which new applications were sought. Among several alternative new application fields, such as chassis of electric cars, rotor blades for wind farms proved to be the most successful ones. A similar case is the yard FR. Fassmer, which also has competences in fibre-reinforced composites and uses this competence for, among others, producing spinners and nacelles for the offshore wind energy industry. Also, the Husum shipyard entered the wind energy business due to the decline in shipbuilding at the end of the 1980s. This did, however, not prevent them from bankruptcy, after which the wind farm part of the company was taken over by Jacobs Energie GmbH in 2000. Out of the merger with pro+pro and BWU, the company was then renamed RePower Systems. However, most yards, such as SSW Schichau Seebeck Shipyard in Bremerhaven (2009), TKMS Nordseewerke in Emden (2010), HDW in Kiel (2010) (Kieler Nachrichten, 2011a) and Nordic Yards in Wismar (2010), have only recently taken this step in relation to offshore wind energy. They are mainly pushed into it because of threatening bankruptcy as a consequence of the negative prospects in shipbuilding. There are several examples in which German shipyards have competed and lost orders to Korean and Polish shipbuilders to build special ships and platforms (see also Zeit, 2010). Hence, the majority of yards entered the new market only after it has taken off already for some time mainly because of push factors (declining prospects in core business) and, regarding Schleswig-Holstein, because of close market observation and learning from entrepreneurial pioneers in Lower Saxony and Bremen.

Of the yards that managed to enter the offshore business, only a part of them produce components in-house. The remaining confine themselves to the renting out of infrastructure (halls, premises, quays, etc.) to producing firms. Constructing offshore towers, as strived for by some yards, is relatively easy than producing other components. Competition, however, is much stronger in this market segment than in other segments, such as rotor blades. Overall, we can observe that of the five large system offshore engineering companies, Siemens Wind Power, Vestas, Bard Group, Areva/Multibrid and RePower Systems, only RePower Systems has some roots in shipbuilding (one of the four companies out of which it has emerged stems from shipbuilding).[3]

Among the main component manufacturers, some emerged out of large companies in other industries diversifying into this new business field. Examples are steel construction (AMBAU Stahl- und Anlagebau, Georgsmarienhütte Holding and, for some time, Thyssen Krupp Stahl Service Center), construction (Züblin AG) and logistics services (Beluga-Hochtief Offshore). Others (Menck, EEW), as mentioned previously, started off by producing hammers and pipes for the offshore oil and gas industries before entering the wind energy market. Since 2000, many of these large companies have been setting up subsidiaries in the coastal regions of Germany to serve the newly emerging market. In addition to these firms, some newly founded firms could establish themselves on the market, particularly in niches of new challenges, such as offshore specification, 5 MW installations (one example is Multibrid as the manufacturer of nacelles). During the last few years, expansion and diversification took place through takeovers of local firms by national and international companies, such as AN Wind Energie taken over by Siemens Wind Power, Multibrid taken over by Areva (France) and RePower taken over by Suzlon (India). This has also led to an increase in investment sums in new offshore wind farm complexes.

Hamburg has an interesting position in the offshore wind energy industry in Northern Germany. Here, we find neither production nor component manufacturers, but instead the offshore wind energy R&D and planning divisions of the large energy concerns (Eon, RWE, Vattenfall, EnBW and GE). To some extent, it can be regarded as the brain port of offshore wind energy in Northern Germany. A specialization within the value chain is also a possibility for Husum in Schleswig-Holstein in order to strengthen its position. The port facilities are, on the one hand, too small for attracting heavy-weight component and pile manufactures. On the other hand, however, the sea access and waterfront infrastructure are argued to be sufficiently large for service and maintenance vessels serving the offshore farms in the North Sea. Some interviewees in Schleswig-Holstein, therefore, see large potentials for their mainland- and island-based ports (e.g. Helgoland and Hörnum) in the niche market of maintenance and repair, although assembly and production of heavy-weight components are also considered in the region's offshore strategy, for example, for Rendsburg and Kiel (Windcomm, 2010). Others, however, are more sceptical, as they point out that competitors in repair and service from the UK, Denmark, Norway and the Netherlands are more experienced, as their markets developed much earlier. Moreover, large maintenance and repair orders are only expected from approximately 2020 onwards, as during the first 6 years wind farm producers are obliged to take care of the maintenance and repair themselves.

All in all, we find relatively few originally newly established firms, if so in the service sector or other fringe areas, which have low entry barriers due to low investment sums required than other parts of the offshore industry. The archetypical spin-off processes leading to a clustering process are thus under-represented in the case of the offshore wind energy industry (Mossig, 2000; Bünstorf & Fornahl, 2009). However, despite an apparent lack of spin-offs, some interviewees are optimistic about future industrial locations in the German regions. Market demands for heavy-weight components such as foundations, piles, towers and transformation platforms are predicted to increase fundamentally, which facilitates shipyards, and others, to potentially enter the wind energy market. SIAG in Emden may be a good example in this respect as it specialized in steelwork and has recently taken on the production of generators and piles.

5. Conclusions

During the last 10 years, the offshore wind energy industry emerged as a new industry in Northern Germany. The industry is still in its early development phase and hence could not yet unfold its full regional economic effects. Nevertheless, the industry is currently already a significant economic sector in the region of Bremen/Bremerhaven and to a lesser extent in the Husum region in Schleswig-Holstein and Rostock in Mecklenburg-Vorpommern.

Since the current production capacities are much lower than the aims set by the European governments concerning renewable energies, in general, and offshore wind energy, in particular, further employment growth can be expected in the future.

The core question of this paper focused on the conditions under which the development path of this industry emerged and the role of the declining shipbuilding industry herein. The overall results show that primarily onshore wind firms or firms of other industries have diversified into offshore wind energy. Only a few offshore wind energy firms have their roots in shipbuilding. The shipbuilding industry and the yard crises only indirectly affected two explanatory factors. First, the access to seaports is a basic locational condition for both offshore wind energy production and shipbuilding. This is a common location factor, not a factor that is transferred from the old to the new industry. Second, the yard crises and shipbuilding decline generated a push factor for companies and policy-makers to look for new market areas. The shipbuilding industry could provide human capital and competences, as well as specific infrastructure, which both could be well used by the new offshore wind energy industry. These effects, however, are of an indirect nature, as there was a time lag of several years between the decline of the shipbuilding industry, on the one hand, and the emergence of the offshore wind energy industry, on the other hand. An important pull factor unrelated to the shipbuilding industry is the long-term stability of the feed-in tariffs at 15 cents/kWh, which has led to strong growth dynamics.

Building on existing regional conditions, policy-makers in Bremen have deliberately tried to change the framework conditions. Further positive signals for the offshore wind energy industry have been the first relocations of firms into the northern region and the extension of scientific expertise in offshore wind energy in the region. The path, therefore, has been created by a combination of one triggering event (positive market development) and favourable regional conditions in Northern Germany, which in turn result from the sea location and the history of the region, on which dedicated actions particularly taken by the Bremen regional policy-makers have built upon. One isolated factor would probably not have led to the development of the offshore wind energy industry in the region. Comparing the regions within Northern Germany, this paper shows that Bremen and Lower Saxony have taken the lead in the emerging offshore wind energy industry, whereas Schleswig-Holstein and Mecklenburg-Vorpommern are following. Why is this the case? The role of agents in search for opportunities, the re-use of resources and the transfer of competences as the basis of new growth (Martin & Sunley, 2010) is essential in explaining these differences. Policy-makers and other institutional agents in Bremen and Lower Saxony have clearly been in support of the industry much earlier than their counterparts in Schleswig-Holstein and Mecklenburg-Vorpommern. Some pessimistic interviewees in the two latter states doubt whether they will be able to catch up with the north-western region in the future. Without strong support, but due to agglomeration advantages, the position of Hamburg as a brain port of the offshore wind energy industry in Northern Germany is exceptional and quite promising at the same time.

Although the shipbuilding industry only had indirect effects on the emergence of this new development path, interestingly, feedback forces work from the new to the old shipbuilding industry. Some yards, for instance, have received new orders to build special ships for offshore wind farms or are currently planning to produce installation, service and maintenance ships. Examples of these yards are the Lloyd Werft in Bremerhaven, Abeking & Rasmussen in Lemwerder, Diedrich Oldersum in Moormerland, Cassens Werft in Emden, the Mützelfeldwerft in Cuxhaven, HDW in Kiel and Nordic Yards in

Wismar. Additional positive effects can be observed if ship maintenance and all other necessary services needed for installing, maintaining and repairing offshore wind farms are located in one home seaport. Particularly in this business area, large hopes and expectations can be seen among policy-makers for positive regional economic multiplier effects for shipbuilders and other maritime service industries. It is particularly in Schleswig-Holstein where big hopes are set on this niche market (Windcomm, 2010).

Thus, shipbuilding only has indirect effects on the emergence of this new path; new impulses are more important as an explanatory factor. However, after the establishment of the offshore wind energy industry, de-locking of shipbuilding activities can be observed. The established companies and regional networks were pushed to reconfigure themselves by several forces, such as new technological, marketing and sales challenges, new network partners and the transplantation of external knowledge into the region. This has had two consequences. First, it has led to a revitalization of the special ship business fields of many yards. Second, other yards which lacked a division of special ships entered straight away into the offshore wind energy business. Both phenomena have slowed down the decline of the shipbuilding development path.

Feeding back to the theoretical concepts used in this paper to explain the emergence of new industries in space, it has become clear that contrasting the windows of locational opportunity with path creation (Martin & Sunley, 2006) has not been very useful in our case. The former very much argues from a footloose, often high-tech industry perspective (Storper & Walker, 1989). Offshore wind energy is obviously not footloose in character, as strong winds and port locations are indispensable. The locational freedom is, therefore, limited to just a few number of locations in Northern Germany. An eclectic approach combining the concept of windows of locational opportunity with path creation is, therefore, more useful to explain the emergence of new industries than a polarizing approach. Concerning path creation, we found the time lag between old and new paths as an important hindering factor that has been neglected somewhat in theoretical elaborations.

The emerging industrial development path is still in its early stages and the industry also faces several risks and challenges with respect to regulation, financing, grid infrastructure and (storage) technologies. Further stable incentive-oriented framework conditions, such as the extension of the EEG law and the recent turnaround in Germany's energy policy after the Japanese nuclear meltdowns in disfavour of nuclear energy (*Economist*, 2011), are essential to the long-term development of the offshore wind energy industry in Northern Germany. However, Germany's strict environmental laws (e.g. regarding the regulation of underwater pile-driving sound, nature and landscape protection) as well as long-term political procedures of permitting wind farms are argued to delay investments. The financial crisis in 2008, for example, has severely threatened the financial security of wind farm projects in Germany, but also led to think about alternative investment strategies such as state guarantees (Staatsbürgschaften) (*Spiegel*, 2010). Moreover, to reach the stage of a stable offshore wind energy cluster, a critical mass of activities and self-reinforcing processes are necessary concerning, for instance, human capital, co-operations and spin-offs, which could lead to a further acceleration of growth (Brenner & Fornahl, 2006). Currently, the regions in Northern Germany, as well as other regions along the North Sea in the UK, the Netherlands, Denmark and Norway, are competing with each other; stable clusters have not been established yet. In order to support the regions in Northern Germany further, more investments in infrastructure, in general, and in offshore terminals, in particular, are necessary. It might make sense to focus this support on the most promising locations in north-western Germany (Bremen and Lower Saxony). However, since also within federal Germany, the northern regional states compete with each other, it is likely that there will be no strong concentration of support for one area. Paradoxically, although many interviewees in Schleswig-Holstein and Mecklenburg-Vorpommern are pessimistic about catching up with the Bremen area in offshore wind energy industry, they have recently set up policy initiatives aiming at emulating the latter region in becoming a strong cluster in offshore wind energy.

Finally, although the determinants of the emergence of the offshore wind energy industry in Northern Germany have been traced back in this paper in a qualitative manner, generalizations on the basis of this explorative research concerning the explanation of the emergence of this industry in other regions are hardly possible. More survey-based quantitative research, therefore, needs to be carried out in the future to get a clearer picture of the current state of the art of the industry and its future potentials. Moreover, comparative research on other coastal regions in neighbouring countries, such as Denmark, Norway, the Netherlands and the UK, is necessary to find out more about the national framework conditions for the emergence of this new industry.

Notes

1. Northern Germany consists of the states of Lower Saxony, Schleswig-Holstein and Mecklenburg-Vorpommern, as well as the city states of Bremen and Hamburg.
2. The costs are estimated at no less than €30 billion, so that from 2020 onwards, large parts of Europe can be provided with electricity powered by wind, solar and wave power works from the North Sea grid (Weser Kurier, 2010a).
3. Germanwind's study of the 40 core companies in the offshore wind energy cluster gave similar results.

References

Arthur, B. W. (1988) Self-reinforcing mechanisms in economics, in: P. W. Anderson, J. A. Kenneth & D. Pines (Eds) *The Economy as an Evolving Complex System: The Proceedings of the Evolutionary Paths of the Global Economy Workshop*, held September 1987, Santa Fe, New Mexico, pp. 9–31 (Redwood City, CA: Addison-Wesley).

Arthur, B. W. (1994) Positive feedbacks in the economy, in: B. W. Arthur (Ed.) *Increasing Returns and Path Dependence in the Economy*, pp. 1–12 (Ann Arbor, MI: The University of Michigan Press).

Bassanini, A. P. & Dosi, G. (2001) When and how chance and human will can twist the arms of clio: An essay on path dependence in a world of irreversibilities, in: R. Garud & P. Karnøe (Eds) *Path Dependence and Creation*, pp. 41–68 (Mahwah, NJ: Lawrence Erlbaum).

Bathelt, H. & Boggs, J. S. (2003) Towards a reconceptualization of regional development paths: Is Leipzig's media cluster a continuation of or a rupture with the past? *Economic Geography*, 79(3), pp. 265–293.

Bathelt, H. & Glückler, J. (2002) Wirtschaftsgeographie in relationaler Perspektive: Das Argument der zweiten Transition, *Geographische Zeitschrift*, 90(1), pp. 20–39.

Bathelt, H. & Glückler, J. (2003) *Wirtschaftsgeographie* (Stuttgart: UTB-Ulmer).

Bierly, P. E., Damanpour, F. & Santoro, M. D. (2009) The application of external knowledge: Organizational conditions for exploration and exploitation, *Journal of Management Studies*, 43(3), pp. 481–509.

Boschma, R. A. (1997) New industries and windows of locational opportunity: A long-term analysis of Belgium, *Erdkunde*, 51(1), pp. 12–22.

Boschma, R. A. (2004) Competitiveness of regions from an evolutionary perspective, *Regional Studies*, 38(9), pp. 1001–1014.

Boschma, R. A. & Frenken, K. (2006) Why is economic geography not an evolutionary science? Towards an evolutionary economic geography, *Journal of Economic Geography*, 6(3), pp. 273–302.

Boschma, R. & Frenken, K. (2011) The emerging empirics of evolutionary economic geography, *Journal of Economic Geography*, 11(2), pp. 295–307.

Boschma, R. A. & Lambooy, J. G. (1999) Evolutionary economics and economic geography, *Journal of Evolutionary Economics*, 9(4), pp. 411–429.

Brenner, T. & Fornahl, D. (2006) Lokale Cluster—Theorie, empirische Erkenntnisse und politische Implikationen, in: R. Reith, R. Pichler & C. Dirninger (Hrsg.) Innovationskultur: Modelle—Indikatoren—Transfer, pp. 185–210 (Innsbruck: Studienverlag).

Bünstorf, G. & Fornahl, D. (2009) B2C—bubble to cluster: The dot.com boom, spin-off entrepreneurship, and regional industry evolution, *Journal of Evolutionary Economics*, 19(3), pp. 349–378.

Cowan, R. & Gunby, P. (1996) Sprayed to death: Path dependence, *Lock-In and Pest Control Strategies, The Economic Journal*, 106(436), pp. 521–542.

Crouch, C. & Farrell, H. (2004) Breaking the path of institutional development? Alternatives to the new determinism, *Rationality and Society*, 16(1), pp. 5–43.

David, P. A. (1993) Historical economics in the longrun: Some implications of path-dependence, in: G. D. Snooks (Ed.) *Historical Analysis in Economics*, pp. 29–40 (London: Routledge).

Dirksmeier, P. (2009) "Don't believe the hype": Kommunale Förderstrategien für die Creative Industries. Das Beispiel der Freien Hansestadt Bremen, *The Planning Review*, 179(4), pp. 37–45.

Dorenkamp, A. & Mossig, I. (2010) Economic policy and its impact on the evolution of clusters and spatial systems exemplified by the German TV program production, in: D. Fornahl, S. Henn & M.-P. Menzel (Eds) *Emerging Clusters. Theoretical, Empirical and Political Perspectives on the Initial Stage of Cluster Evolution*, pp. 43–73 (Cheltenham: Edward Elgar).

Economist (2011) German energy; nuclear? Nein, danke, June 2.

Eich-Born, M. (2005) Schiffbau in Europa im Zeitalter der Globalisierung, *Geographische Rundschau*, 57(12), pp. 54–61.

Eich-Born, M. & Hassink, R. (2005) On the battle between shipbuilding regions in Germany and South Korea, *Environment and Planning A*, 37(4), pp. 635–656.

Frenken, K., van Oort, F. & Verburg, T. (2007) Related variety, unrelated variety and regional economic growth, *Regional Studies*, 41(5), pp. 685–697.

Garud, R. & Karnøe, P. (2001) Path creation as a process of mindful deviation, in: R. Garud & P. Karnøe (Eds) *Path Dependence and Creation*, pp. 1–38 (Mahwah, NJ: Lawrence Erlbaum).

Garud, R., Kumaraswamy, A. & Karnøe, P. (2010) Path dependence or path creation? *Journal of Management Studies*, 47(4), pp. 760–774.

Giese, E., Mossig, I. & Schröder, H. (2011) *Globalisierung der Wirtschaft. Eine wirtschaftsgeographische Einführung* (Paderborn: Schöningh).

Grabher, G. (1993) The weakness of strong ties: The 'lock-in' of regional development in the Ruhr area, in: G. Grabher (Ed.) *The Embedded Firm: On the Socio-Economics of Industrial Networks*, pp. 255–277 (London: Routledge).

Haller, F., Landsberg, H. & Wehling, W. (2003) *Für eine Tourismus-Initiative*, Bremen/Bremerhaven—Bilanz, Ziele, Projektideen, Wirkungen, BAW-Monatsbericht 4.

Hassink, R. (2005) How to unlock regional economies from path dependency? From learning region to learning cluster, *European Planning Studies*, 13(4), pp. 521–535.

Hassink, R. (2010) Locked in decline? On the role of regional lock-ins in old industrial areas, in: R. Boschma & R. Martin (Eds) *Handbook of Evolutionary Economic Geography*, pp. 450–468 (Cheltenham: Edward Elgar).

Hassink, R. & Shin, D.-H. (2005) South Korea's shipbuilding industry: From a couple of cathedrals in the desert to an innovative cluster, *Asian Journal of Technology Innovation*, 13(2), pp. 133–155.

Hirsch, P. M. & Gillespie, J. J. (2001) Unpacking path dependence: Differential valuations accorded history across disciplines, in: R. Garud & P. Karnøe (Eds) *Path Dependence and Creation*, pp. 69–90 (Mahwah, NJ: Lawrence Erlbaum).

IWR (2008) *Zur Struktur der Windenergieforschung in Deutschland. Untersuchung der deutschen Wind-Forschungsstrukturen im internationalen Vergleich mit Handlungsempfehlungen an die Politik. Studie im Auftrag des Bundesumweltministeriums* (Münster: Internationales Wirtschaftsforum Regenerative Energien).

Jovanović, M. N. (2009) *Evolutionary Economic Geography: Location of Production and the European Union* (London: Routledge).

Karnøe, P. & Garud, R. (1995) Path creation and dependence in the danish wind turbine field. Working paper, Institute of Organization, and Industrial Sociology. Copenhagen Business School.

Kern, H. & Schumann, M. (1990) *Rationalisierung in der industriellen Produktion. 4. Auflage*, 4th ed. (München: C.H. Beck).

Kieler Nachrichten (2010) Offshore: Kiel hinkt hinterher, Kieler Nachrichten, press release, July 14.

Kieler Nachrichten (2011a) Auf HDW kommt jede Menge Arbeit zu, Kieler Nachrichten, press release, March 22.

Kieler Nachrichten (2011b) Nordseehäfen stecken ihren Kurs ab, Kieler Nachrichten, press release, March 9.

Kramm, K. (1980) *Die Seeschiffswerften und ihre Standorte in der Bundesrepublik Deutschland unter Berücksichtigung der Zulieferindustrie und der internationalen Wettbewerbslage* (Essen: Tosch).

MacKinnon, D., Cumbers, A., Pike, A., Birch, K. & McMaster, R. (2009) Evolution in economic geography: Institutions, Political Economy, and Adaptation, *Economic Geography*, 85(2), pp. 129–150.

Martin, R. (2010) Roepke lecture in economic geography—rethinking regional path dependence: Beyond lock-in to evolution, *Economic Geography*, 86(1), pp. 1–28.

Martin, R. & Sunley, P. (2006) Path dependence and regional economic evolution, *Journal of Economic Geography*, 6(4), pp. 395–437.

Martin, R. & Sunley, P. (2010) The place of path dependence in an evolutionary perspective on the economic landscape, in: R. Boschma & R. Martin (Eds) *Handbook of Evolutionary Economic Geography*, pp. 62–92 (Cheltenham: Edward Elgar).

Mossig, I. (2000) Lokale Spin-off-Gründungen als Ursache räumlicher Branchencluster. Das Beispiel der deutschen Verpackungsmaschinenbau-Industrie, *Geographische Zeitschrift*, 88(3/4), pp. 220–233.

Mossig, I. (2006) *Netzwerke der Kulturökonomie: Lokale Knoten und globale Verflechtungen der Film- und Fernsehindustrie in Deutschland und den USA* (Bielefeld: Transcript-Verlag).

Mossig, I., Fornahl, D. & Schröder, H. (2010) Heureka oder Phoenix aus der Asche? Der Entwicklungspfad der Offshore-Windenergieindustrie in Nordwestdeutschland, *Zeitschrift für Wirtschaftsgeographie*, 54(3–4), pp. 222–237.

Musgrove, P. (2010) *Wind Power* (Cambridge: Cambridge University Press).

North, D. C. (2005) *Understanding the Process of Economic Change* (Princeton: Princeton University Press).

Nuhn, H. (1990) Schiffbau in Norddeutschland—Krise und Strukturwandel eines traditionellen Industriezweigs an der Küste, *Mitteilungen der Geographischen Gesellschaft in Hamburg*, 80, pp. 341–366.

Nuhn, H. (1998) Maritime Wirtschaft in Norddeutschland, in: E. Kulke (Hrsg.) *Wirtschaftsgeographie Deutschlands*, pp. 309–343 (Stuttgart: Klett-Perthes).

Phene, A. & Almeida, P. (2008) Innovation in multinational subsidiaries: The role of knowledge assimilation and subsidiary capabilities, *Journal of International Business Studies*, 39(5), pp. 901–919.

Rao, H. & Singh, J. V. (2001) The construction of new paths: Institution-building activity in the early automobile and biotechnology industries, in: R. Garud & P. Karnøe (Eds) *Path Dependence and Creation*, pp. 243–267 (Mahwah: Lawrence Erlbaum).

Rigby, D. L. & Essletzbichler, J. (1997) Evolution, process variety, and regional trajectories of technological change in U.S. manufacturing, *Economic Geography*, 73(3), pp. 269–284.

Schamp, E. W. (2000) *Vernetzte Produktion. Industriegeographie aus institutioneller Perspektive* (Darmstadt: Wissenschaftliche Buchgesellschaft).

Seebeck, M. (2011) Auswirkungen politischer Rahmenbedingungen auf die Entstehung und Entwicklung eines Offshore-Windenergie-Clusters im Nordwesten Deutschlands, unpublished MA thesis, University of Kiel.

Song, J., Almeida, P. & Wu, G. (2003) Learning-by-hiring: When is mobility more likely to facilitate interfirm knowledge transfer? *Management Science*, 40(4), pp. 351–365.

Spiegel (2010) Offshore-Ausbau: Röttgen drückt bei Hochsee-Windparks aufs Tempo, press release, July 28.

Storper, M. (2011) Why do regions develop and change? The challenge for geography and economics, *Journal of Economic Geography*, 11(2), pp. 333–346.

Storper, M. & Walker, R. (1989) *The Capitalist Imperative. Territory, Technology, and Industrial Growth* (New York: Basil Blackwell).

Tholen, J. & Ludwig, T. (2005) *Beschäftigung, Auftragslage und Perspektiven im deutschen Schiffbau. Ergebnisse einer Betriebsrätebefragung im September 2005*, IAW Forschungsbericht 9, Oktober 2005, Universität Bremen.

Van den Bosch, F. A. J., Volberda, H. W., Boer, M. & de (1999) Coevolution of firm absorptive capacity and knowledge environment: Organizational forms and combinative capabilities, *Organization Science*, 10(5), pp. 551–568.

VSM (Verband für Schiffbau und Meerestechnik e.V.) (2010) Jahresbericht 2009. Hamburg.

Weser Kurier (2010a) Neues Stromnetz in der Nordsee. Anrainerstaaten wollen Milliarden in die Nutzung erneuerbarer Energien investieren, press release, January 6.

Weser Kurier (2010b) Tauziehen um neuen Offshore-Hafen, press release, January 26.
Windcomm (2010) Offshore-Strategie, Handlungsempfehlungen für Schleswig-Holstein (Husum: Windcomm).
Zahra, S. A. & George, G. (2002) Absorptive capacity: A review, reconceptualization and extension, *Academy of Management Review*, 27(2), pp. 185–203.
Zeit (2010) Windenergie-Boom lässt deutsche Werften hoffen, press release, October 29.

Spatial Processes of Industry Emergence: US Wind Turbine Manufacturing Industry

GREGORY THEYEL

Institute for Manufacturing, University of Cambridge, 17 Charles Babbage Road, Cambridge CB3 OFS, UK

Abstract *This research seeks to understand the underlying locational resources and dynamics explaining the location and growth of new, environmental technology industries. Beginning with the location patterns of the US wind turbine manufacturing industry, this research uses demand, supply and policy factors to explain how the resources and dynamics of a location help shape its future industrial composition. The findings include the prominent role of demand factors, including wind power resources, existing installed wind power and electricity prices as well as supply factors, but the weakness of the influence of governmental policy. This research can assist companies, investors and policy-makers understand the spatial dynamics of industry emergence in order to better align their strategies and policies.*

1. Introduction

Demand, supply and policy factors can influence firm location and the spatial emergence of new industries. Companies choose locations and industries emerge where demand conditions are strong, when interaction with customers is important or when transportation delivery costs are high (Malmberg & Maskell, 2002; Boschma & Wenting, 2007; Cooke & Ehret, 2009). Proximity to suppliers can be important for innovation and supply chain management (Feldman & Audretsch, 1999; Breschi & Lissoni, 2001; Boschma, 2004). National, state and local governments frequently attempt to attract companies and grow new industries through policy mechanisms (Hung & Chu, 2006; Aldrich & Fiol, 2007; Block & Keller, 2010).

Researchers have written a great deal about agglomeration and to a lesser extent about the spatial emergence of industries. Very little has been written about the location decisions of companies and industry emergence and agglomerations in environmental technology industries (for exceptions see Cooke, 2008 and Russo, 2003) where demand, supply and policy mechanisms influence the location choices of these highly desirable industries. Little is known about the spatial patterns of these emerging industries and the influence of different forces on the location choices of companies in these industries.

The purpose of this paper is to help understand how demand, supply and policy conditions affect the location of newly emerging environmental technology industries, as demonstrated by the US wind turbine manufacturing industry. The paper is organized as follows: Section 2 reviews prior research on location and emerging industries and formulates hypotheses about the determinants of firm location and agglomeration in the emerging US wind turbine manufacturing industry. Section 3 describes the methods and introduces the data employed for the analysis. Section 4 describes the findings from the empirical analysis. Section 5 presents conclusions and implications.

2. Literature: Spatial Resources and Dynamics and Emerging Industries

Industries are often unevenly spread due to increasing returns to scale, differences in market access and locational specificity (Feldman & Francis, 2004). Companies often locate where there is sophisticated demand, specialized suppliers, firms in related industries and supportive government policies (Porter, 2003). In addition, locations usually possess legacy resources, which shape their future industrial composition (Klepper, 2002; Rosenthal & Strange, 2003; Martin & Sunley, 2006). Industries can emerge from older, established industries. These foundational industries can offer new technology, spin-offs and support for emerging industries (Bresnahan *et al.*, 2001; Boschma & Frenken, 2009; Neffke *et al.*, 2011). Locations with these attributes can offer competitive advantages for the firms located there such as direct feedback and assistance from demanding customers, a supportive policy environment, labour force pooling, inter-firm collaboration, supplier networks and technology. Location-specific resources can enhance technological innovation and create opportunities that encourage entrepreneurship and the emergence of new industries (Boschma & Wenting, 2007). This locational entrepreneurship is described by Boschma and Frenken (2009) as a "branching process" where new opportunities are built from the resources and capabilities of existing regional industries and institutions.

Locations can offer firms competitive advantages if they possess certain characteristics. Technology development, knowledge creation and entrepreneurship are said to be aided by demand, supply and policy characteristics (Feldman & Francis, 2004; Scott, 2006; Lorentzen, 2008; Malecki, 2010). For example, the presence of sophisticated customers can help firms with insight on customer needs, more rapid feedback leading to innovations and an extension of firms' research and development in the form of user innovations and co-development of products and processes (von Hippel, 2005). Locations with concentrations of related companies (suppliers, complementors and competitors) can access a network of human talent, participate in teaming and even benefit from rivalry that spurs innovation and entrepreneurship (Marshall, 1890; Torre & Rallet, 2005; Maskell & Malmberg, 2007). In addition, regions can offer supporting institutions such as government policy, universities, incubators and industry associations (Morgan, 2007). These locational aspects may help explain why economic development is not the same in all locations. The following offers more on spatial resources and dynamics specific to the wind turbine industry in the USA. This leads to hypotheses that may help explain the spatial patterns of companies in this industry.

2.1 *Demand Factors*

Proximity to demand may be an important determinant for the location of wind turbine manufacturing companies. Companies located in proximity to their customers may be better able to learn about their customers' needs and even co-develop knowledge through the exchange of products and services (Maskell & Malmberg, 2007). Customer requirements may spur innovation, and it is likely that proximity to demanding customers can enhance the learning process for both parties (Porter, 2003). Companies working with local customers can gain traditional benefits of collaboration such as risk and resource sharing, but they can also improve their communication and information flow with these local customers (Feldman & Francis, 2004).

A location's wind power resources, its installed capacity of wind power and its electricity prices are possible demand signals for wind turbine manufacturing companies. Wind power resources are not the same everywhere, but spread out unevenly across the USA. Wind resources exist in vast quantities across the Great Plains, on mountain ridges and offshore (US Department of Energy, 2010). One might assume that resources determine the location of renewable energy generation because without resources, generation is not possible. Often the development of renewable resources for energy production follows the concentrations of the renewable energy resources. However, the distribution of generated power beyond close proximity locations requires large-scale distribution and transmission lines that are dependent on high investments (Russo, 2003). Wind turbine manufacturers may be drawn to wind power resources and installed wind turbines in order to assist with new development due to the heavy nature of the equipment and the subsequent high transportation costs to ship equipment and to service existing installations (Russo, 2003). Customers may be wind farm developers, electric utilities and even farmers and industrialists. The price of electricity may also affect the opportunities for wind power generation. While wind generated power is usually one of the cheapest types of renewable power, it still lags behind hydro- and natural gas-generated electricity. Therefore, wind turbine manufacturers may benefit from being near installed wind turbines and renewable energy resources and in states with high electricity prices. These demand factors can be formulated into hypotheses to test their influence on the location of wind turbine manufacturing companies.

Hypothesis 1a: Availability of wind power resources attracts wind turbine manufacturing companies.
Hypothesis 1b: Installed wind power attracts wind turbine manufacturing companies.
Hypothesis 1c: High electricity prices attract wind turbine manufacturing companies.

2.2 *Supply Factors*

The industrial composition of a location may be a valuable resource for the evolution of new industries. This "new path creation" (Martin & Sunley, 2006; Martin, 2010) can occur as new industries develop from the expertise, experiences and extensions of old industries (Neffke *et al.*, 2009). Companies in these older industries become suppliers and sources of employees, spin-offs, investment capital and innovations for companies in the new industry (Boschma & Wenting, 2007). In addition, a higher degree of variety among similar industries appears to be best for knowledge creation, innovation and growth (Frenken *et al.*, 2007; Neffke *et al.*, 2011).

Wind turbine manufacturing is supported by companies in industries such as iron foundries, metal fabrication, motors and generators, electronic equipment and power transmissions (US Department of Energy, 2010). Turbine companies are likely to gain quicker knowledge of evolving information and technology and availability of equipment by being near these supply industries. This proximity may help wind turbine manufacturers know their options when they are addressing a problem or developing new technology. These local companies can be a source of knowledge or help develop new knowledge, and proximity between firms and supportive companies can quicken this process, lower costs and promote the exchange of information, the building of trust and the pooling of resources to jointly design new specialized equipment and products. This may explain the location patterns of wind turbine manufacturing companies and will be tested with the following hypothesis.

Hypothesis 2: Wind turbine manufacturers locate in proximity to supply industries.

2.3 *Policy Factors*

The role of government during the emergence of an industry tends to change as the technology and companies mature. Governments use a range of mechanisms to influence the direction and pace of industry growth, and they tend to use them at different times for different effects (Hung & Chu, 2006). These include: grants, subsidies and procurement; intellectual property protection; investment incentives; regulation/standards; employment incentives; and macroeconomic policies (Block & Keller, 2010).

In the period before an industry begins to truly emerge, and in the earlier stages of its existence, government's role is most likely to involve support for basic research with limited emphasis on seeding particular new industries, i.e. technology discovery. General areas of research are supported with the focus on ways to address technical challenges or need to expand the application of existing technologies. Government support in the form of grants, subsidies and procurement contracts is also often important. The government may continue these roles as an industry evolves, but interventions may be more targeted, at both product and process improvements, and their roles in the establishment of regulations and the setting of standards are likely to become more prevalent as the risks and opportunities of industries are better understood (Aldrich & Fiol, 2007) and as competition increases.

Government policy has been shown to be the primary driver for the adoption of renewable energy (Haar & Theyel, 2006). State governments in the USA have been seeking ways to encourage the generation of renewable energy, especially due to the growing concerns about climate change and dependence on foreign oil reserves. Government policy can encourage the development of renewable energy with regulations, mandates and incentives. The government can open competition, which can encourage the development of new energy resources and technologies. Governments can mandate the adoption of renewable energy with energy portfolio standards. This government policy action (deregulation, portfolio standards and grants for adopting renewable energy technology) may spur demand for renewable energy development and technology such as wind turbines. This leads to the following hypothesis:

> Hypothesis 3: State-level energy policies such as deregulation, renewable energy portfolio standards and grants attract wind turbine manufacturing companies.

3. Data and Methods

Environmental technology can be categorized as a "product, service or process that delivers value using limited or zero non-renewable resources and/or creates significantly less waste than conventional offerings" (Pernick & Wilder, 2007, p. 2). Environmental technology industries include equipment needed for cleaning operations such as air and water purification, recycling and technology for the generation of renewable energy, such as solar and wind power. This paper focuses on the US wind turbine manufacturing industry, which has grown significantly from 1998 to 2008.

There were 206 wind turbine manufacturing companies in the USA in 2008 (US Census Bureau, 2010). This is a more than 100% increase in the number of companies since there were 98 in 1998. Figure 1 and Table 1 show the uneven distribution of the companies across the USA. California, Texas and Pennsylvania have the most companies, and Connecticut, Delaware and South Carolina have the highest location quotients in 2008 (US Census Bureau, 2010). Table 1 also shows the change in the state location quotients

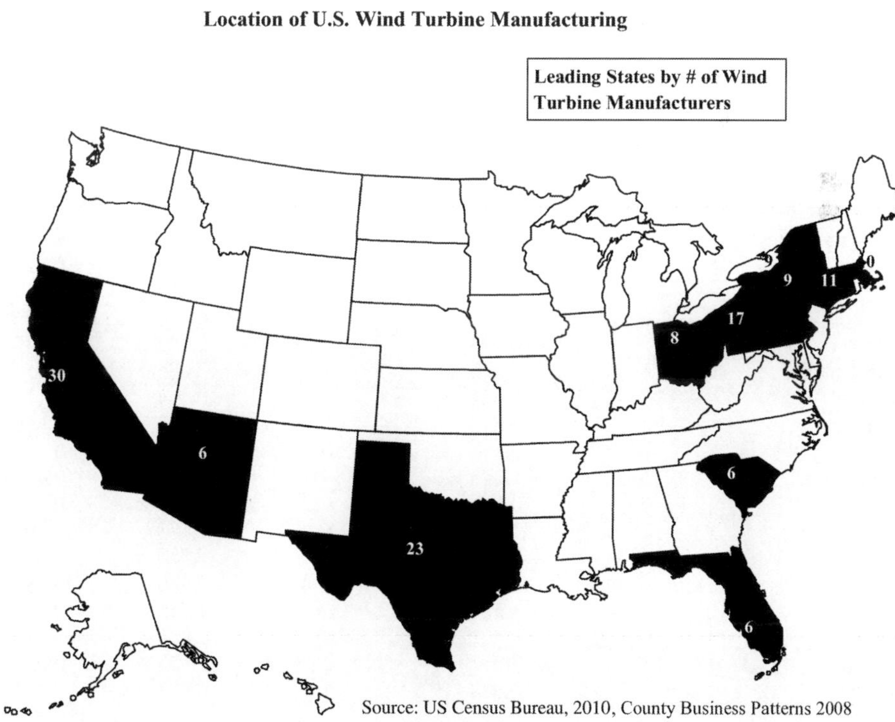

Figure 1. Location of US wind turbine manufacturing.

Table 1. US wind turbine manufacturers

State	Wind turbine manufacturers (1998)	State location quotient (1998)	Wind turbine manufacturers (2008)	State location quotient (2008)	1998–2008 location quotient change
Alabama	1	0.681	2	0.643	−0.038
Alaska	0	0	0	0	0
Arizona	4	2.990	6	1.915	−1.076
Arkansas	0	0	1	0.525	0.525
California	9	0.672	30	1.098	0.426
Colorado	0	0	5	1.514	1.514
Connecticut	5	3.220	11	3.607	0.387
Delaware	0	0	1	2.457	2.457
District of Col.	0	0	0	0	0
Florida	4	0.934	6	0.683	−0.250
Georgia	0	0	0	0	0
Hawaii	0	0	0	0	0
Idaho	2	4.527	0	0	4.527
Illinois	1	0.208	4	0.412	0.204
Indiana	6	2.389	4	0.714	−1.675
Iowa	1	0.969	5	2.068	1.099
Kansas	0	0	2	0.998	0.998
Kentucky	0	0	2	0.771	0.771
Louisiana	1	1.044	0	0	−1.044
Maine	1	1.990	2	1.785	−0.205
Maryland	0	0	0	0	0
Massachusetts	9	3.525	10	2.093	−1.432
Michigan	2	0.465	5	0.593	0.128
Minnesota	3	1.371	5	1.031	−0.340
Mississippi	1	1.251	1	0.617	−0.634
Missouri	1	0.495	2	0.469	−0.025
Montana	0	0	0	0	0
Nebraska	0	0	0	0	0
Nevada	1	2.198	4	3.266	1.068
New Hampshire	0	0	1	0.771	0.771
New Jersey	3	0.952	3	0.530	−0.422
New Mexico	0	0	1	1.017	1.017
New York	9	1.414	9	0.780	−0.634
North Carolina	3	0.993	4	0.644	−0.349
North Dakota	1	5.242	1	2.085	−3.156
Ohio	3	0.621	8	0.794	0.173
Oklahoma	4	3.647	5	2.017	−1.630
Oregon	0	0	0	0	0
Pennsylvania	6	1.299	17	1.773	0.474
Rhode Island	0	0	0	0	0
South Carolina	2	1.666	6	2.171	0.505
South Dakota	0	0	1	1.510	1.510
Tennessee	0	0	2	0.480	0.480
Texas	9	1.516	23	1.748	0.232
Utah	0	0	4	1.876	1.876
Vermont	1	3.011	2	2.950	−0.062
Virginia	1	0.620	3	0.833	0.213

(Continued)

Table 1. Continued

State	Wind turbine manufacturers (1998)	State location quotient (1998)	Wind turbine manufacturers (2008)	State location quotient (2008)	1998–2008 location quotient change
Washington	1	0.475	3	0.628	0.153
West Virginia	0	0	0	0	0
Wisconsin	3	1.108	5	0.826	−0.283
Wyoming	0	0	0	0	0
Total	98		206		

Source: US Census Bureau 2010, County Business Patterns 2008.

between 1998 and 2008. A location quotient is a measure of the concentration of an industry in a location. For this research, it is the measure of state-level concentration compared to the entire USA. The uneven distribution of wind turbine companies and the increasing concentration of companies (more than half of the states have a negative change in location quotient from 1998 to 2008 or have no companies) suggest there are factors driving the agglomeration of this emerging industry.

The hypotheses presented in Section 2 are tested using data collected from secondary sources including the US Census Bureau, US Department of Energy and the Database of State Incentives for Renewables and Efficiency (DSIRE). The renewable energy and policy data were collected for 2008, and the state industry data were collected for 1998 and 2008. The 50 US states and the District of Columbia were considered individual locations for this research. While this is not optimal because economic activity may not follow state lines, company location data and energy resource, generation, price and policy data were only available for this unit of analysis. This is less problematic than it may appear as, based on the author's analysis, over 90% of the 50 largest metropolitan locations are largely contained within the borders of one state, with notable exceptions being the New York/New Jersey and Pennsylvania/New Jersey economic concentrations.

In order to test the hypotheses posed above, and, in turn, better understand the underlying locational resources and dynamics explaining the location and growth of the wind turbine manufacturing industry, variables were developed for each hypothesis to be analysed with the firm locations in 2008. Table 2 shows the dependent variable (whether the firm is located in a state with a higher portion of wind turbine manufacturers than the US in general, i.e. a location quotient greater than 1.0) and the independent variables relevant to the data. Independent variables representing demand factors include the state wind generation potential in megawatts, as presented by the US Department of Energy (2010), the installed wind power capacity by state (representing an installed base of potential customers for wind turbine companies) and the state price of electricity, which, if relatively high, could attract wind power development, and thus offer opportunities for wind turbine manufacturers. One independent variable representing supply industries is the combined state location quotient for five supply industries that provide inputs for the manufacture of wind turbines (see Figure 2). There are three independent variables—all measured as a 1 or 0 depending on whether the state does or does not have the policy measure in place by 2006—for assessing the influence of policy factors on the location

Table 2. Data and sources

Characteristic	Source	Hypothesis
Clustered state manufacturer (CSM)	US Census Bureau *County Business Patterns*, 2010	Dependent variable
State wind power potential (WINDRES)	US Department of Energy, *Wind Powering America*, 2010	Demand
Installed wind power capacity (INSTALLWIND CAP)	US Department of Energy, *Wind Powering America*, 2010	Demand
Price of electricity per kilowatt (ELECT$)	US Department of Energy, *Wind Powering America*, 2010	Demand
Supply industries location quotients (SUPINDUST)	US Census Bureau *County Business Patterns*, 2010	Supply
Deregulation (DEREG)	Database of State Incentives for Renewables and Efficiency, 2010	Policy
Renewable portfolio standards (RPS)	Database of State Incentives for Renewables and Efficiency, 2010	Policy
Grants (GRANTS)	Database of State Incentives for Renewables and Efficiency, 2010	Policy

of wind turbine manufacturing. These include state-level deregulation, a renewable portfolio standard (RPS), which mandates that state utilities have a percentage of renewable energy in their generation portfolios by a set date, and a grants programme for encouraging the development of renewable energy generation.

This study uses Spearman correlation (Table 3) and logistic regression (Table 4) analysis of the factors to isolate the effects of the variables identified above (see Table 2) on firm location. The specific focus of the correlations and logistic regression model is the overall influence of demand, supply and policy factors on a firm's location in a clustered state (i.e. a state with a higher portion of wind turbine manufacturers than the US in general). A challenge of this research was addressing the direction of cause and effect between the dependent and independent variables.

Longitudinal data made it at least partially possible to ascertain whether associated factors existed prior to the locating of wind turbine manufacturers. Policy data are from approximately 2005 (DSIRE, 2010), as are data on electricity prices (US Department of Energy, 2010), installed wind power capacity and the location of supply industries, while the turbine manufacturing location data are from 2008 (US Census Bureau, 2010). To this effect, the logistic regression analysis uses factors only present prior to the locating of a large proportion of the wind turbine manufacturers. The following

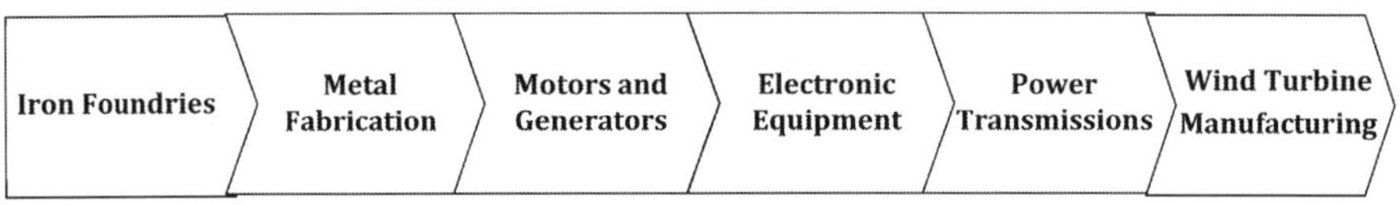

Figure 2. Wind turbine manufacturing supply chain.

Table 3. Spearman correlation results

	Observations	Mean	SD	Min.	Max.	(1)	(2)	(3)	(4)	(5)	(6)	(7)
(1) Clustered state manufacturer	206	0.690	0.464	0.000	1.000							
(2) Wind potential	206	28017	46323	0	138400	0.288**						
(3) Installed wind power capacity	206	1907	2886	0	9410	0.344**	0.457					
(4) Electricity price	206	7.440	1.810	4.030	11.930	0.092	−0.128	−0.026				
(5) Supply industries	206	1.017	0.139	0.750	1.460	0.460**	−0.226	−0.387**	−0.303**			
(6) Deregulation	206	0.660	0.475	0.000	1.000	−0.171	0.259**	0.311**	0.416**	−0.192**		
(7) Renewable portfolio standards	206	0.906	0.292	0.000	1.000	0.077	0.246**	0.282**	0.444**	−0.207**	0.234**	
(8) Grants	206	0.645	0.479	0.000	1.000	−0.186**	0.105	0.031	0.077	−0.317**	−0.075	0.221**

*Significant level of 5%.
**Significant level of 1%.

Table 4. Logistic regression results

	Model (1)
Wind potential	0.000 (0.000)*
Installed wind power capacity	0.001 (0.000)***
Electricity price	0.565 (0.140)***
Supply industries	5.177 (1.625)*
Deregulation	−0.400 (0.431)
Renewable portfolio standards	0.035 (0.681)
Grants	−2.072 (0.507)***
Nagelkerke R^2	0.418
Observations	206
Constant	−8.172 (2.248)***

Notes: Dependent variable: location in clustered state. The table contains coefficient estimates; figures in parenthesis are standard errors. Nagelkerke R Square offers a pseudo R^2 statistic.
*Significant level 10%
**Significant level 5%
***Significant level 1%

equation organizes the logistic regression analysis.

$$\begin{aligned} p(\text{Clustered State Manufacturer} = 1) &= f(\beta 1\text{WINDPOTENT} + \beta 2\text{INSTALLWINDCAP} \\ &\quad + \beta 3\text{ELECT\$} + \beta 4\text{SUPPINDUST} + \beta 5\text{DEREG} \\ &\quad + \beta 6\text{RPS} + \beta 7\text{GRANTS}) + \varepsilon \end{aligned}$$

Different location-specific factors influence the location choices of wind turbine manufacturers. The following section presents the results of the correlations and logistic regression model showing the influence of factors affecting the location of companies in states with clusters of wind turbine manufacturers.

4. Results: Spatial Resources and Dynamics for Industry Emergence

Supply and some demand factors play prominent roles in the location of wind turbine manufacturers, but natural resources and the influence of governmental policy are less important. Spearman correlation (Table 3) shows association and logistic regression analysis (Table 4) shows the positive influence of both demand and supply factors on the location of wind turbine manufacturers. Surprisingly, the policy factors show either negative association or no influence on the location of turbine makers. Analysis of the correlation results (all <0.5) and proxy tests of the regression variables using a linear regression model with Tolerance and VIF showed that all of the independent variables in the logistic regression model met the limits in tests for multicolinearity. The following sections offer detail on the research findings.

4.1 *Demand for Wind Turbine Manufacturers*

There is a significant association and influence of wind power potential (resources) and wind power installed capacity on the location of wind turbine manufacturers, but the association is quite strong, while the influence is slight as indicated by the minuscule (0.000 and 0.001) coefficient values for both variables in the regression model. The opposite is the case for electricity prices, which are not significantly associated with turbine manufacturer location, but are significantly influential based on the regression results. These results are support for all three demand hypotheses (1a, 1b and 1c), though to different extents, and these findings seem to indicate that these three factors combine to explain firm location choices in this context. This is likely to indicate that the quality of wind energy resources and proximity of existing customers and even higher electricity prices are fairly weak individually, but together they represent a context for the location of wind turbine manufacturers. In addition, the installed capacity of wind power is likely to represent local support for wind power, which is not captured in other variables (Russo, 2003).

4.2 *Supply for Wind Turbine Manufacturers*

Wind turbine manufacturing companies are not located uniformly across the USA (as discussed above and shown in Figure 1 and Table 1). Wind turbine companies are clustered with each other and with supply industries. The data analysis shows support for Hypotheses 2 that turbine companies are located in proximity to supply industries, such as iron foundries, metal fabrication, motors and generators, electronic equipment and power transmissions (see Figure 2). Supply industries exhibit the highest coefficient in the regression model, and this analysis shows that turbine manufacturers are likely to benefit from greater availability of needed inputs and access to insight and technology development for their own innovation. Wind turbine companies are likely to benefit from proximity to supportive companies, which allows for cost-reducing and information-sharing benefits of economies of agglomeration. This offers support for assertions from Neffke *et al.* (2009, 2011) that a new industry is more likely to be established in a region when it is related in terms of skills and technologies to the already existing regional industrial base (Henning *et al.*, 2010; Martin, 2010).

4.3 *Policy for Wind Turbine Manufacturers*

State governments have made efforts to influence their energy markets. The array of the indirect and direct policy measures to support renewable energy development shows the substantial interest in the successful expansion of renewable energy (DSIRE, 2010). However, this research shows that renewable portfolio standards are not significant correlation or regression indicators of the location of wind turbine manufacturing companies (see Tables 3 and 4). Renewable portfolio standards require utilities to generate a set percentage of electricity from renewable sources (DSIRE, 2010). In most cases, this policy measure has been put in place only recently or is currently being phased in. This may be why there is no significant statistical relationship between this policy and the location of wind turbine companies. There is also no significant regression relationship and a negative correlation with deregulation of state energy markets, which allows for competition among generators of electricity and the location of wind turbine companies. Most wind installations require significant up-front capital, which precludes small energy developers. Therefore, electric utilities and large-scale developers are most likely to install wind power, and deregulation is likely not to motivate them to install wind power, and hence, as discussed above, lead to the attraction of wind turbine manufacturers. The relationship between grants for developing renewable energy (for utilities and private parties, depending on the state) is negative as shown for correlations and the regression model. It is possible the grant programmes are either too new and/or small to have an effect on attracting wind turbine manufacturers. These regression results do not lend support for Hypotheses 3a, 3b or 3c, and based on these measures, state energy policy does not influence the location of wind turbine manufacturing companies.

5. Conclusions and Implications

This research shows how the emerging wind turbine manufacturing industry is spatially agglomerated and becoming increasingly concentrated. Demand and supply factors, which represent the industrial legacy of a location, are associated with and influence the location of wind turbine manufacturing companies. Proximity to wind resources, installed wind power generation and supply industries are statistically significant indicators of agglomerations of wind turbine manufacturing companies. These findings contribute to our understanding of firm location patterns in an environmental technology industry, and that environmental technology industries may be difficult to "seed" with policy measures. Instead, environmental technology industries, like wind turbine manufacturing, may emerge and concentrate in locations with a legacy of demand and supply industries. This section suggests trends and implications of this research for investors, environmental technology entrepreneurs, their suppliers and customers and policy-makers.

Investors, start-up companies and corporate entrepreneurs may benefit from this research by using the results to better understand the location and growth of companies in new industries, namely, environmental technology. This understanding could help these entities with their investment decisions as companies in locations with optimal demand and supply conditions may be more likely to grow and prosper.

Companies may use the results of this research to help with their strategies and location decisions. Suppliers can offer wind turbine manufacturers innovations and seek locations close to these companies so that they can work more closely to develop applications for location-specific challenges. Turbine manufacturers can benefit from locating close to their suppliers so that they can receive more responsive service and extend their capabilities through inter-firm collaboration.

This research can also help policy-makers understand industry patterns as they seek to balance energy policy with economic development. Wind turbine companies follow demand (wind power potential, installed wind power capacity and electricity prices); hence, policy-makers should focus on attracting the wind turbine manufacturing industry if their location has the relevant demand and supply resources and dynamics. If their location lacks these resources and dynamics, it is likely to be more effective for them to focus on different industries that are a better match with their particular locational context.

Finally, this research contributes to our understanding of the underlying resources and dynamics explaining the location and growth of emerging industries. In particular, this research may help with predicting where environmental technology industries are likely to emerge based on regions' industrial demand and supply legacies. While wind turbine manufacturers may be seeded from demand conditions and supply industries, other types of environmental technology may be seeded in other locations from different industrial legacies. Future research can help illuminate these questions about which regional legacies are more likely to generate which new industries.

References

Aldrich, H. & Fiol, M. (2007) Fools rush in? The institutional context of industry creation, in: A. Cuervo, D. Ribeiro & S. Roig (Eds) *Entrepreneurship Concepts, Theory and Perspective* (New York, NY: Springer Publishing).

Block, F. & Keller, M. (Eds) (2010) *State of Innovation: The U.S. Government's Role in Technology Development* (Boulder, CO: Paradigm Publishers).

Boschma, R. (2004) Competitiveness of regions from an evolutionary perspective, *Regional Studies*, 39(9), pp. 1001–1014.

Boschma, R. & Frenken, K. (2009) Technological relatedness and regional branching, in: H. Bathelt, M. Feldman & D. Kogler (Eds) *Dynamic Geographies of Knowledge Creation and Innovation* (New York, NY: Routledge).

Boschma, R. & Wenting, R. (2007) The spatial evolution of the British automobile industry: Does location matter? *Industrial and Corporate Change*, 16(2), pp. 213–238.

Breschi, S. & Lissoni, F. (2001) Knowledge spillovers and local innovation systems: A critical survey, *Industrial and Corporate Change*, 10(4), pp. 975–1005.

Bresnahan, T., Gambardella, A. & Saxenian, A. (2001) Old economy inputs for new economy outcomes: Cluster formation in the new silicon valleys, *Industrial and Corporate Change*, 10(4), pp. 835–860.

Cooke, P. (2008) Regional innovation systems, clean technology, and Jacobian cluster-platform policies, *Regional Science Policy and Practice*, 1(1), pp. 23–45.

Cooke, P. & Ehret, O. (2009) Proximity and procurement: A study of agglomeration in the Welsh aerospace industry, *European Planning Studies*, 17(4), pp. 549–567.

Database of State Incentives for Renewables and Efficiency (DSIRE) (2010) Available at http:// www.dsire.com (accessed 10 December 2010).

Feldman, M. & Audretsch, D. (1999) Innovation in cities: Science-based diversity, specialization and localized competition, *European Economic Review*, 43(2), pp. 409–429.

Feldman, M. & Francis, J. (2004) Homegrown solutions: Fostering cluster formation, *Economic Development Quarterly*, 18(2), pp. 127–137.

Frenken, K., Van Oort, F. & Verburg, T. (2007) Related variety, unrelated variety, and regional economic growth, *Regional Studies*, 41(5), pp. 635.

Haar, N. & Theyel, G. (2006) U.S. electric utilities and renewable energy: Drivers for adoption, *International Journal of Green Energy*, 3(3), pp. 271–281.

Henning, G., Stam, E. & Wenting, R. (2010) Path Dependence and Path Creation in Regional Economic Development working paper, Utrecht University, Utrecht.

von Hippel, E. (2005) *Democratizing Innovation* (Cambridge, MA: MIT Press).

Hung, S. & Chu, Y. (2006) Stimulating new industries from emerging technologies: Challenges for the public sector, *Technovation*, 26(1), pp. 104–110.

Klepper, S. (2002) The Evolution of the US Automobile Industry and Detroit as Its Capital. The Ninth Congress of the International Schumpeter Society, March 28–30, 2002.

Lorentzen, A. (2008) Knowledge networks in local and global space, *Entrepreneurship and Regional Development*, 20(6), pp. 533–545.

Malecki, E. (2010) Everywhere? The geography of knowledge, *Journal of Regional Science*, 50(1), pp. 493–513.

Malmberg, A. & Maskell, P. (2002) The elusive concept of localization economies: Towards a knowledge-based theory of spatial clustering, *Environment and Planning A*, 34(3), pp. 429–449.

Marshall, A. (1890) *Principles of Economics* (London: MacMillan).

Martin, R. (2010) Rethinking regional path dependence: Beyond lock-in to evolution, *Economic Geography*, 86(1), pp. 1–27.

Martin, R. & Sunley, P. (2006) Path dependence and regional economic evolution, *Journal of Economic Geography*, 6(4), pp. 395–437.
Maskell, P. & Malmberg, A. (2007) Myopia, knowledge development and cluster evolution, *Journal of Economic Geography*, 7(5), pp. 603–618.
Morgan, K. (2007) The learning region: Institutions, innovation and regional renewal, *Regional Studies*, 21(1), pp. 147–159.
Neffke, F., Svensson Henning, M., Boschma, R., Lundquist, K. J. & Olander, L. O. (2009) The Dynamics of Agglomeration Externalities along the Life Cycle of Industries working paper, Utrecht University, Utrecht.
Neffke, F., Henning, M. & Boschma, R. (2011) How do regions diversify over time? Industry relatedness and the development of new growth paths in regions, *Economic Geography*, 87(3), pp. 237–265.
Pernick, R. & Wilder, C. (2007) *The Clean Tech Revolution* (New York, NY: HarperCollins Publishing).
Porter, M. (2003) The economic performance of regions, *Regional Studies*, 37(6&7), pp. 549–578.
Rosenthal, S. & Strange, W. (2003) Geography, industrial organization, and agglomeration, *Review of Economics and Statistics*, 85(2), pp. 377–393.
Russo, V. M. (2003) The emergence of sustainable industries: Building on natural capital, *Strategic Management Journal*, 24(4), pp. 317–331.
Scott, A. (2006) Entrepreneurship, innovation and industrial development, *Small Business Economics*, 26(1), pp. 1–24.
Torre, A. & Rallet, A. (2005) Proximity and localization, *Regional Studies*, 39(1), pp. 47–59.
US Census Bureau (2010) *County Business Patterns*. Available at http://www.census.gov/econ/cbp (accessed 10 December 2010).
US Department of Energy (2010). *Wind Powering America*. Available at http://www.eere.energy.gov (accessed 10 December 2010).

Clean Energy Technology and the Role of Non-Carbon Price-Based Policy: An Evolutionary Economics Perspective

NICHOLAS HOWARTH

Environmental Change Institute, School of Geography and the Environment, University of Oxford, UK

ABSTRACT *Much academic attention has been paid to the role of carbon pricing in developing a market-led response to low-carbon energy innovation. Taking an evolutionary economics perspective, this paper makes the case as to why price mechanisms alone are insufficient to support new energy technologies coming to market. In doing so, we set out the unique investment barriers in the clean energy space. For guidance on possible approaches to non-carbon price-based policies that seek to tackle these barriers, we turn to case studies from Asia, a region which has experienced a strong uptake in climate policy in recent years.*

1. Introduction

There is an argument emerging among economists that a carbon pricing mechanism alone is insufficient to transform the carbon-intensive base of the global economy (Acemoglu *et al.*, forthcoming; Stern, 2009). Although a transparent and robust carbon price (set through a carbon tax or emissions trading scheme) is vital for changing the production decisions of firms at the margin in favour of low-carbon-intensive substitutes, the deployment of clean technologies into the market requires additional government intervention through non-price-based policies (also called "technology" or "innovation" policies).

In this paper, we focus on those technology policies which would operate in conjunction with a carbon price to reduce barriers to private investment in clean technologies. These non-price fiscal policies, we argue, work best when they are targeted at overcoming early-stage market failures. Without these early-stage barriers being sufficiently addressed, the introduction of a carbon price may be pre-mature in the sense of creating non-optimal social outcomes. This is because these barriers, which are independent of a carbon price, prevent new technologies from entering the market and competing with fossil fuel sources of energy generation.

The dual role for carbon pricing and technology policy in light of the multiple market failures at play has been recognized by Lord Stern and others. As the 2006 Stern Review on the Economics of Climate Change argues,

> [T]he presence of a range of other market failures and barriers mean [*sic*] that carbon pricing alone is not sufficient . . . Technology policy, the second element of a climate change strategy, is vital to bring forward the range of low-carbon and high-efficiency technologies that will be needed to make deep emissions cuts.

There is international agreement that private investment will need to provide the bulk of financing for renewable energy projects globally (Doornbosch & Knight, 2008). Given that we are yet to see private investment meet the scale and scope need for new clean energy investment in both developed and developing countries, governments may seek to introduce fiscal policies of the nature described in this paper.

Each sector and each country will face its own unique investment barriers for the financing and deployment of new technologies. In this paper, we take examples from the Asia Pacific Economic Cooperation (APEC) region to examine the possible technology policy options. APEC is a worthwhile region to examine as it contains many of the world's major emitters and has been a focus of major climate policy innovation over the last decade.

2. Role of Evolutionary Economics in Climate Policy

The traditional economic analysis of climate change has most often been grounded in the theory of externalities inherent in the standard welfare economics of Pigou (1912, 1932). This focuses on establishing the costs and benefits of an emissions target or level of pollution and the optimal or most cost-effective way to achieve it through the regulation of greenhouse gas (GHG) prices, quantities or both. To do this, two broad alternative approaches are usually considered: the Pigouvian tax, where polluters pay for the externality they impose on others, and the Coasian approach, which allocates property rights for using the environment (Coase, 1960) and which, due to its political economy, has been termed more of a pollutee-pays approach (Helm, 2005). Here, Helm argues that the process underpinning the allocation of pollution rights under the Coasian approach allows polluters to recognize the property right as an asset and pass on the cost of meeting emissions targets, as exemplified in the "windfall profits" generated by energy companies upon the introduction of the European Union's Emissions Trading Scheme. This means that the pollutee (the consumer) bears the cost, rather than the polluter (the energy provider). To balance each approach's various disadvantages and advantages, hybrid schemes have also been investigated. For a selection of works in this area, see Baumol and Oates (1988), Cropper and Oates (1992), Pizer (2002), Jacoby and Ellerman (2004), Philibert (2009), Hepburn (2006), Nordhaus (2007), Fankhauser and Helpburn (2010a, 2010b), Brohé *et al.* (2009) .

The logic of the standard approach follows that having "internalized the externality of pollution" through establishing a price on carbon, the problem of energy policy is reduced to ensuring that energy prices reflect the full social costs of energy production and utilization. This is on the basis that the world is far too complex for politicians to "pick winners" and, aside from providing some informational support or changing relative prices to correct for any externalities, it is best to leave people "free to choose" and let "the market decide" as much as possible.

Market-driven interventions, such as carbon price-based policy, fit comfortably within the rubric of neoclassical equilibrium analysis, where well-informed consumers purchase products and energy services in a way that maximizes their own welfare and, by extension, best promotes the interests of society at large. It was Hayek in the 1930s who championed this idea over that of more centrally planned regulatory approaches which he observed emerging in the Union of Soviet Socialist Republics. He argued that in the market system, actors can be coordinated through a single signal—prices—whereas, in a planned system, to efficiently allocate resources across the economy, regulators would need to know all the utility and product functions of all the actors—an impossibly vast amount of information. Hayek argued that this uncertainty would create "government failure" where, through acting on incomplete information or becoming beholden to the special interests of a powerful minority, the interests of society at large would be compromised. In the 1960s, this notion was further supported by applications of economic logic to the political process directly through public choice theory, providing yet another warning of the imperfections of centrally planned regulations (Olson, 1965). Within the economic mainstream, this helped support a culture which argued that while market failures were bad, intervention to correct them should be approached very cautiously. These concerns were further heightened given that outside of the limited theoretical setting of economic modelling exercises, the actual economy was thought to be ridden so profoundly with market failures. Even if a policy-maker managed to put in place an efficient policy with respect to, say, a carbon tax, it could actually lead to a fall in aggregate welfare because of interactions with market failures elsewhere in the economy (see Corden, 1997[1974]).

This view of the government regulation, while still influential today, perhaps had its high point with the Reagan–Thatcher years of the 1980s, which was defined by quips of economic management such as "The nine most terrifying words in the English language are: 'I'm from the government, and I'm here to help'". and "Governments bad. Markets good".

It would be wrong, however, to characterize the economic mainstream as seeing low-carbon technological change as something to be completely determined through the broad interaction of price signals in the market for a completely homogenous good—energy. For example, Nicholas Stern argues in his Review on the Economics of Climate Change that

> Many commentators are skeptical about technology policy, saying it is wrong for bureaucrats to "pick winners". There is something in this, but it is also naïve or dogmatic in its underlying assumption that markets work perfectly unless distorted by government. In this case, markets do not work well unless assisted by government. (p. 111).

The theory of induced innovation, for example, is a hypothesis that seeks to explain the direction of the technological change underpinning invention in the traditional model through the impact of factor prices (Ahmad, 1966; Kamien & Schwartz, 1968; Binswanger, 1974). Within this body of work, technological externalities have been identified, providing a rationale for government intervention in a specific technology market (Acemoglu *et al.*, forthcoming). For example, technological externalities in research and development (R&D) are likely to exist because many technologies are non-rival (one person's use of the technology need not exclude another's) and because they are non-excludable, meaning that property rights with respects to the technology are difficult to enforce. Because firms are assumed to be profit maximizers, they will only invest in R&D if they can capture sufficient economic rents from the effort. As this investment is likely to have broader applicability and can benefit the rest of the economy, positive spillovers are likely to exist. Isaac Newton's "standing on the shoulders of giants" is an apt metaphor here. The conclusions from this body of research for energy, the environment and innovation are usefully surmised by Grubb and Ulph (2002, p. 104):

> ...while environmental policies [such as carbon pricing] may induce innovation that will lead to cleaner technologies, the theoretical and empirical evidence we have does not give us a great deal of confidence that environmental policies alone will be sufficient to bring about major environmental innovation. It seems, that to have a significant impact, it will be necessary to pursue both environmental and technology policies. Given that we are dealing with two distinct areas of externality [the GHG externality and the technology externality], this conclusion is hardly surprising.

Porter and van der Linde's (1995) work has been influential in challenging the orthodox economic view (Palmer *et al.*, 1995) by highlighting the positive role that environmental regulations can play in promoting environmentally beneficial innovation and supporting economic competitiveness at the firm, sector and nation-state levels. They argue that tighter environmental standards can not only reduce business costs directly, but also spur cost-reducing innovation further, thus boosting competitiveness. This has given rise to the so-called Porter hypothesis, which has inspired a large body of supportive empirical studies (e.g. Costantini & Crespi, 2007; Horbach, 2007; Lanoie *et al.*, 2008; Kriechel & Ziesemer, 2009). However, other economists have rejected such case-study-based approaches by arguing that the examples used are special cases. It is just as likely that environmental regulations come at a net cost across the economy and embody a significant opportunity cost. For example, Jorgenson and Wilcoxen (1990) and Hazilla and Kopp (1990) used a dynamic general equilibrium model to show that such regulations are necessarily cost adding because of the manner in which environmental regulations depress other "productive" investments. The Porter hypothesis remains a highly controversial issue with price theorists strongly arguing that innovation is driven by changing the price of factors of production (see Newell *et al.*, 1999).

One of the great advantages of the traditional economic analysis is its parsimony. However, as is often the way, in its greatest strength lies greatest potential weakness. Because the problem of climate change and GHG reduction is modelled in a demand-and-supply framework, the tools at the analyst's disposal are those within the model's analytical logic. As such, focus is on the properties of price and quantity. Their elasticities with respect to demand and supply and marginal change are explored through the application of differential calculus.

However, in this framework, non-price regulation is analytically problematic, so in many cases, it is simply ignored or dismissed as an inferior tool on the basis of the criticisms already discussed. What is lost by the traditional model is a more detailed understanding of the mechanics of economic change. For many short-term or sector-specific circumstances, this may not be problematic. But for analysing long-term, non-marginal phenomena—such as what underpins low-carbon technological transition—it represents a significant theoretical weakness. This is especially the case in industries characterized by increasing returns and where investments have large sunk costs, which can lead to path dependence such as in energy supply and distribution. It is here that this paper argues that an evolutionary economics perspective of climate policy is particularly helpful.

The role of an evolutionary perspective versus the traditional approach for climate policy can be most easily understood through an interrogation of the core metaphors from which each paradigm draws its analytical logic: evolution versus equilibrium. The former emphasizes change, movement, progression, unpredictability, crisis and metamorphosis, whereas the latter emphasizes stability or stasis, a state of rest or balance due to the equal action of opposing forces. As one might expect, the implications of their epistemological differences in economics and public policy are profound.

The assumptions which underlie the fundamental theorems of welfare economics state that in a perfectly competitive market, an equilibrium reached by trading between buyers and sellers at market prices will be economically efficient and that, ceteris paribus, such an equilibrium can always be reached from any given starting allocation of resources (Arrow & Debreu, 1954). This logic means that government interventions are framed in terms of trying to recreate a theoretical model of perfection completion—where the goal is achieving the single equilibrium that allocates resources efficiently across the economy (Foxton, 2011, p. 136). At this point, there is no further impetus for change emerging from within the model. If change does occur, it is through exogenous factors determined outside the model, such as a change in technology or in consumer preferences—and then firms adapt to the new set of circumstances and reach a new equilibrium. While it is acknowledged that these exogenous forces can be profound, the model's logic requires these assumptions to be made to maintain parsimony.

In contrast to the study of equilibrium, the core focus of evolutionary economics is in understanding the processes of change and stability in the economy. A key text in this approach is Nelson and Winter's (1982) Evolutionary Theory of Economic Change. This builds on the work of Schumpeter (1975[1942]) and Simon (1955, 1957) to place technology at the centre of study in the economic model, where economic agents display "bounded rationality", that is, they are limited by their ability to access and process information and hence look for satisfactory "satisficing" solutions. Economic change in this model can come in a variety of forms, at "critical junctures" or in "gales of creative destruction" which can be far from marginal—however, the key point is that change is incessant:

> Capitalism, then, is by nature a form or method of economic change and not only never is, but never can be stationary. And this evolutionary character of the capitalist process is not merely due to the fact that economic life goes on in a social and natural environment which changes and by its change alters the data of economic action; this fact is important and these changes (wars, revolutions, and so on) often condition industrial change, but they are not its prime movers. Nor is this evolutionary character due to a quasi-autocratic increase in population and capital or to the vag-

> aries of monetary systems of which exactly the same thing holds true. The fundamental impulse that sets and keeps the capitalist engine in motion comes from the new consumers' goods, the new methods of production or transportation, the new markets, the new forms of industrial organisation that capitalist enterprise creates. (Schumpeter, 1975[1942], pp. 82–83])

From this, Nelson and Winter (1982) introduced the fundamental notion of the "routine", which could be any technical, procedural, organizational or strategic process used by a firm as part of its normal business activities, for example, its R&D strategy or a particular production profile. They argued that firms compete by searching for better techniques or processes that satisfy their chosen criteria—whether it be for profit, market share or some other objective—and which evolve out of a historically embedded adaptive process. From this, stable routines may emerge, formed by a dominant set of actors. However, this outcome is also highly contingent on the starting point of the system, and so outcomes may be quite heterogeneous according to the geography of the system.

Here, the notion of path dependence—how decisions taken today or in the past can influence future decisions, closing off certain options and opening up others—has been particularly influential (Arthur, 1988, 1989, 1994; David, 1993, 1985, 2005). This body of work emphasizes how increasing returns effect phenomena such as technical interrelatedness, economies of scale, dynamic learning and coordination effects and self-reinforcing expectations, combined with the quasi-irreversibility of investments and large fixed setup costs, which lead to the lock-in of—in our case: high carbon—industrial infrastructure and consumer patterns (Unruh, 2000). Such effects mean that because many high-carbon investment decisions have been made in the past, similar, supportive high-carbon investments are more economically efficient (at least in the short term not taking into account the costs of pollution) in the present when compared with low-carbon alternatives which may embody a different set of technologies, scientific and education resources, legislative changes, and political actors and allegiances.

While the impact of path dependence theory has had a profound effect across the social sciences, it has also been criticized, in its canonical form, as placing too much emphasis on the concept of lock-in to a stable equilibrium state and neglecting de-locking phenomena (Martin, 2010). Thus, the model can describe the emergence of heterogeneous economic structures across different geographies, but once an equilibrium state is "selected", little insight is provided into how the process of change occurs between stable states. This concept, that the economy lurches from one stable state to another, is similar to Schumpeter's "gales of creative destruction" and has been explored within the field of complex systems theory under the umbrella term "punctuated equilibrium" (Beinhocker, 2007).

In this paper, we base our approach on the strategic niche management at its related multi-level perspective (MLP), which arose out of the sociology of technology and was originally developed to understand transitions and regime shifts (Kemp, 1994; Schot *et al.*, 1994; Kemp *et al.*, 1998, 2001; Levinthal, 1998; Rip & Kemp, 1998; Geels, 2002, 2005; Schot & Geels, 2008). As Geels (2002, p. 1259) pointed out, the MLP is not meant as an ontological description of reality, but as a set of analytical and heuristic concepts to understand the mechanics of socio-technical change. Within this framework, Geels and Kemp (2006, pp. 234–236) highlighted three different types of changes according to their scope and underlying mechanisms. First, reproduction only involves change at the socio-technical regime level, not at a landscape or niche level. It involves the reproduc-

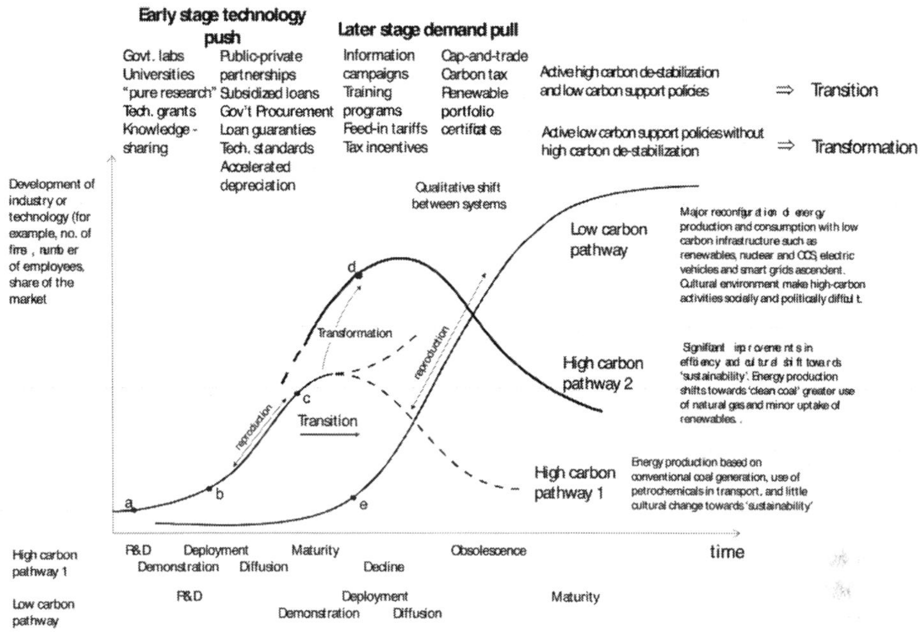

Figure 1. Climate policy and technology: an evolutionary economics perspective.

tion and refinement of existing technological routines. Such innovations can still lead to considerable productivity improvements over time, but there is little re-organization of dominant actors, key technologies or the knowledge base (call this high-carbon pathway 1 in Figure 1). Transformation is change at the regime and landscape levels, but with little interaction from new technological niches. An example could be where a significant change in politics shifts institutional incentives away to a new "cleaner" technological trajectory (such as the imposition of carbon price), but new technologies do not compete to replace the existing systems (high-carbon pathway 2). Finally, a transition is said to occur when interactions between niches, the dominant socio-technical regime and the landscape bring about a major qualitative shift in the economic system. Incumbents are likely to attempt to suppress the new niches which may be better aligned with the social and economic landscape but are yet to achieve market maturity from emerging. If successful, transition implies a significant shift in the knowledge base, new technological objects, infrastructure and regulatory framework along with shifts in consumer preferences, driven forward by new social groups and businesses.

We distinguish between two broad types of climate change policies: technology push, which focuses on protection and support on new niches, and market pull, which change incentives within the dominant socio-technical regime (Burer & Wustenhagen, 2009).

There is growing recognition that it is not possible to rank the cost-effectiveness of all climate change policies in general because the circumstances will change based on the economic settings within a particular country. Furthermore, because of the different geographical contexts in which market failures and path dependence take place, there is growing recognition that social outcomes may be optimized when a mixture of both policy approaches is used strategically (Popp, 2006; Fisher, 2009; Acemoglu *et al.*, forthcoming). For example, economic modelling by Acemoglu *et al.* (2009) shows that relying on a carbon price alone leads to an excessive drop in consumption in the short term as clean energy alternatives are slow to compete in the market. There is also evidence that price-based mechanisms are more suited to marginal efficiency improvement within a specific technological paradigm, shifting the diffusion curve A up and to the left in Figure 1, and not well targeted at delivering paradigm represented by the shift from A to B.

In addition to applying a mix of policies, policy-makers may need to pay attention to the timing and duration of policy options. On the issue of staging, given that technology policy is often focused on early-stage technologies, it is important to introduce this measure early on so that new technologies can emerge quickly. If carbon pricing is introduced in the absence of these "stage-one" policies, then there is a risk that carbon price inflation will be borne fully by electricity consumers because of a lack of viable energy alternatives. However, the strategic staging of these policies can ensure that a strong pipeline of new clean energy technologies will emerge to compete with coal-fired electricity once a carbon price arrives.

On the issue of duration, well-targeted non-price-based policy should aim to be temporary and be reduced once the market failure it is addressing abates and diffusion is underway. If measures are applied indefinitely, then they may reduce the flexibility of the socio-technological system to adapt to changing priorities by reinforcing locking in certain technologies as well as causing an unnecessary drain on national budgets. Thus, an evolutionary response emphasizes the destabilization of the existing high-carbon regime as much as the support of new low-carbon niches.

3. What Are the Investment Barriers?

In this section, we examine in greater detail the early-stage market failures for private investment in clean energy generation technologies. These failures are independent of a price on carbon.

From economic theory, private investors face two principal hurdles when making investment in new technologies. The first hurdle is the "adoption hurdle". Because investors are uncertain about the adoption of a new technology in the market, they are unable to accurately predict the future cash flows from their investment. This leads them to under-invest in new ideas and deploy capital into less risky projects where cash flows are more predictable.

The second hurdle is the "spillover hurdle". Even if the investors are successful in selling a new technology into the market, they are not fully compensated for the public good or positive externalities derived from R&D investment in that new idea. This is because the spillover effects of this investment are captured by competitors who indirectly benefit by replicating the intellectual property and take advantage of new technological know-how. The presence of this incentive asymmetry creates a second barrier to private investment in new technology (Knight, 2010).

The market failures in financing new technologies exist at some level across all technology sectors including the biotechnology and digital communication sectors. In many countries, venture capitalists typically have the highest risk appetite for new technology deals and play an important role in financing ideas out of universities and national laboratories and bringing them to the market. However, there are a number of characteristics about renewable energy technologies which make the above-mentioned market failures more acute than in other sectors.

The first characteristic is that renewable energy projects are very capital intensive compared with biotechnology and IT projects. The upfront cost of a renewable energy technology demonstration project is upwards of US$100 million per project (Ministry of Finance Green Paper, 2009). This is typically too big for venture capitalists who have relatively small funds under management (approximately US$100–500 million) (Shellenberger *et al.*, 2008). In addition, renewable energy projects are competing with biotechnology and digital communications technologies, which cost a fraction of this to demonstrate. By contrast, project financiers and commercial banks which would typically lend capital for deals of this size are highly risk averse and unwilling to take on the technology risk involved. They will finance less risky deals (hospitals, toll roads and airports) ahead of renewable energy projects because the risk/return premium is more predictable and favourable. The result is that the "adoption hurdle" described above is more acute for renewable energy projects because private investors are required to make larger size risks in the sector.

The second characteristic of renewable energy projects which is unique is that the end product—electricity—is homogenous and directly substitutable with coal-fired electricity. This means that unlike new drugs or software applications which may earn a premium in the market, renewable energy competes on price with coal-fired electricity. Given that it is difficult for renewable energy to reach price parity in its production costs with coal-fired electricity, a price-directed intervention is required by the government to guarantee an end market for renewable energy generators. This also amplifies the "adoption hurdle" because investors have no sense of the future cash flows that they may receive from a new technology.

Clearly, a carbon price has an important role in changing the end market for renewable energy. However, a carbon price is technology agnostic and will favour later-stage technologies which have already reached the commercial scale. For early-stage technologies, however, alternative fiscal support through technology-specific feed-in tariffs, direct investment, tax exemptions, accelerated depreciation rates and renewable energy standards (RESs) help reduce the capital costs faced by an investor and guarantee an end market for the product. Without these measures in addition to a carbon price, new renewable energy technologies may struggle to compete with more established renewable energy technologies to reach some economies of scale.

The above analysis relates to the investment barriers faced by technology investment in the clean energy sector. There are two additional points worth mentioning.

The first is that technology financing is not just a challenge for developed economies. Investors in renewable energy projects in developing economies will face a high technology risk even if the technology is imported. This is because the yield which a particular renewable energy technology is able to achieve directly is related to the local environmental conditions. Therefore, the same technology will have a different yield in Jakarta, Indonesia, as it would in Guangzhou, China. Also, different technologies may be better suited to the conditions in different geographies (Knight, 2010). Leaving this to one side, however, a close analysis of clean technology patent data indicates that a number of APEC economies are becoming leading net exporters of clean technologies. Among these, the US, Japan, China and South Korea are most prominent (Dechezleprêtre *et al.*, 2008). Indeed, China's recent strategic document Medium- and Long-term Development Plan for Renewable Energy in China (2007) explicitly identifies the deployment of Chinese intellectual property domestically as a future policy objective.

The second point to be mentioned is that in addition to the market failures identified above, investors in APEC developing economies face barriers to project finance related to the governance profile of these countries. These will be discussed briefly in Section 6. These barriers are not exclusive to the energy sector, but to effectively attract private investment, fiscal policy must be targeted to address these issues.

4. Supply Side (Technology Push)

Supply-side measures refer to policies where the government financially supports the entry of a particular technology into the market. This would be recorded as an expenditure on the government budget. Traditionally, this has been criticized as the "government picking winners". This may be a problem is because the government may be motivated in its selection by criteria other than cost-effectiveness (Popp, 2010). The private sector, at least in a well-governed system, has more transparent incentives to make these kinds of judgements.

There may be reasons other than cost-effectiveness where the government has a strong case to offer supply-side fiscal assistance to a particular technology. One reason might be the competitive advantages which a particular country has in a natural resource—such as geothermal resources in the case of Indonesia. If this exists but the technologies to utilize this advantage are immature and a long way from commercialization, a government may consider subsidizing research and developing funding into certain technologies directly. It is difficult to model the efficiency of public spending if this approach is taken, and R&D funding offers no guarantee that the innovations will be adopted in the market (Yang & Oppenheimer, 2007). This suggests that R&D policies should be adopted with caution. Politically, however, this approach is often desirable because it may have positive spillover effects for job creation and industry development in the local economy.

A more structural reason for government intervention is the high upfront capital costs of renewable energy technologies compared with their return on investment. The cost base for renewable energies is front-loaded with high capital investment and relatively low operating costs (Ministry of Finance Green Paper, 2009). The main hurdle for renewable energy developers is finding bank finance to meet the upfront costs given the level of technology, operational and other project-related risks (Box 1).

Box 1. Financing Indonesia's geothermal assets

Indonesia is host to 40% of the world's geothermal resources, which is more than what any other country in the world can lay claim to. However, only 3% of Indonesia's geothermal capacity is developed.

The Indonesian Green Paper (2009) identifies a number of investment barriers to commercializing Indonesia's geothermal resources. One barrier identified is the large upfront investment costs for geothermal projects combined with uncertainty about the value of each research. Because early drilling and exploration costs are high, private investors are unwilling to take on the risk of pre-tender field surveys and exploration studies. The report recommends a national Clean Technology Fund be set up to bear the initial costs for funding confirmation drilling.

If there is an agreement that high upfront costs are a structural barrier to renewable energy investment, this is not a licence to promote fiscal stimulus for any technologies. Some fiscal policies are arguably better targeted than others at addressing the investment barriers around high upfront costs. Examining current supply-side policies within the APEC region, it is possible to distinguish between those policies targeted at the investment costs of clean energy projects and those targeted at the production costs of energy generation.

4.1 *Policies for Investment Costs*

4.1.1 *Investment tax credit.* Fiscal policies targeting investment costs are arguably best geared to address the high start-up costs of renewable energy technologies. A number of countries have identified high upfront costs in renewable energy projects as an area for fiscal reform. For example, the US has set a federal 30% investment tax credit, which will be maintained through until 2016 for solar photovoltaics (PV), solar thermal power, solar hot water, small wind and geothermal resources. This tax credit is explicitly broad enough to cover utility companies. By providing a time limit on access to this credit, the government is attempting to avoid "lock-in" where new technologies come to depend on the tax credit over the long run.

4.1.2 *Accelerated depreciation.* In India, the government has allowed accelerated depreciation for a number of key technologies. For example, small hydro and wind projects are allowed 80% accelerated depreciation on equipment and devices used in the first year of installation of the project. In the case of biomass power generation, there is 100% deprecation in the first year of the project for many of the key large-scale components including boilers and waste recovery equipment.

The purpose of accelerated depreciation is to reduce taxable income in current years in exchange for increased taxable income in future years. This encourages businesses to purchase new assets upfront and helps with the initial costs of starting up these major projects.

4.1.3 *Tax holidays.* In China, the high upfront costs of clean energy projects are reduced by tax holidays for selective taxes which contribute to initial project costs. These include import duties which affect imported renewable energy technologies and value-added tax. The tax holidays applied to renewable energy technologies in China are reported in Box 2.

Box 2. China's climate finance tax incentives

The Chinese government is currently offering a number of tax incentives to reduce the upfront costs of deploying renewable energy projects. In China, the standard value-added tax is 17%. However, for a selection of renewable technologies, the VAT is reduced. It is only 13% for biogas, 8.5% for wind and 6% for small hydro projects.

Import duties are also lower for renewable energy technologies (many technologies are imported). Average import duty is currently 23%, but for renewable energy, these rates are significantly lower. It is 3% for the components of wind power plants, 6% for wind turbines and 12% for PV systems).[1]

4.1.4 *Loan schemes.* It might be possible to offer loan guarantees to project developers who struggle to attract bank finance for large renewable energy projects. While there is a risk that these policies might result in underwriting bad loans to poorly conceived projects, there will also be instances where projects are indeed bankable but local financial service professionals lack the experience or risk appetite to complete these deals. The risks of the former may be mitigated by private–public co-financing models where both parties are financially exposed to default losses. This will help address some of the teething problems around the capital-intensive nature of clean energy projects where the financial requirements are disproportionately high for the amount of project risk involved.

In the US, a number of conditional loan guarantees were offered by the Department of Energy to promising venture-backed clean technology companies which were struggling to raise capital. Under the Energy Policy Act of 2005, sectors affected by these loan guarantees included a number of capital-intensive clean technologies such as biomass, solar, wind, hydropower, carbon sequestration and advanced fossil energy coal. Beneficiaries included Bright Source and Solyndra. The multilateral development banks have also positioned themselves to play a supportive role with regard to major infrastructure project finance in developing countries (see Box 3).

Box 3. Publicly backed guarantees for infrastructure

A number of multilateral development banks have sought to deploy public finance to address the capital intensity challenge of renewable energy projects with high levels of technology risk. The presence of public sector bank guarantees or subordinate loans can assist with the commercial terms of project finance for new renewable energy projects. In the absence of these interventions, private investors may be unwilling to finance a project.[2]

The European Investment Bank offers equity financing and financial guarantees for selected large-scale infrastructure schemes. Examples of tools used include guarantees for pre-completion or early operational risk, subordinated loans or mezzanine finance.

The World Bank and Global Environment Fund have set up the Geothermal Energy Development Fund. It offers partial risk guarantees for risks such as the short-term upfront geological risk of exploration.

At a local level, a similar approach may be taken for homeowners or building occupants seeking to finance climate-related installations. Mexico, for example, has established "green mortgages" for homeowners seeking to finance solar hot water installations into their houses at favourable interest rates. In the Philippines, the United Nations Environment Programme and the Global Environment Fund have supported the Solar Home Systems Financing Programme in Palawan. This programme offers publicly backed financial security to households seeking to purchase solar home systems. The security comes in the form of a loss reserve fund which offers assistance in the event of default.

4.2 *Policies for Production/Operating Costs*

A number of countries have begun to introduce fiscal policies to reduce the production costs of renewable energy generation. For example, the American Recovery and Reinvestment Act (2009) extended the eligibility of companies generating wind, solar, geothermal and "closed-loop" bioenergy for production tax credits (until 2013, and 2012 for wind). These tax credits last for the first 10 years of the facility's operation.

In Indonesia, the government has also sought to introduce policies to tackle the production cost of renewable energy generation. Legislation has recently been put in place to subsidize the production costs of small-scale renewable energy systems (Ministerial Decree No. 1122/K/30/MEM/2002 on Small Distributed Power Generation Using Renewable Energy and Ministerial Regulation No. 2/2006 on Medium-Scale Power Generation Using Renewable Energy). The intention has been to offset the distortion in the market caused from subsidized conventional energy generation which makes it difficult for renewable energy to compete in the market.

However, it has been suggested that this Indonesian legislation has not been effective in attracting new investment into the market. As the Ministry of Finance's Green Paper on economic and fiscal policy strategies for climate change mitigation in Indonesia comments, "these production costs [of small-scale renewable energy plants] are not transparent, long, and difficult". Furthermore, the Green Paper points out that "negotiations between the investor and utility are required to reach an agreement that will provide the necessary return for the investor". However, the state utility, Perusahaan Listrik Negara, has been unwilling to come to an agreement with the government because it regards renewable energy as unprofitable.

The situation in Indonesia suggests that this policy approach does not sufficiently address the underlying investment barriers in the development and deployment of renewable energy. It is more important for financiers and utility companies considering investment in new, risky technologies to have a long-term stable cash flow. Providing security on this issue is arguably better addressed through demand-side measures which create more certainty for demand of renewable energy in the market. These will be considered in the next section.

5. Demand Side (Technology Pull)

One of the main investment barriers to new renewable energy capacity is the lack of certainty which private investors have in the end market for clean technologies (the adoption hurdle). Without a clear expectation of demand for their product or service and therefore future revenue streams, private investors will typically under-invest in R&D or not take on the financing risks of commercial-scale demonstration.

In the case of energy, renewable energy generation supplies a product (electricity) which is directly substitutable with fossil fuel-based energy generation. In most cases, renewable energy technologies are not at the stage where their production costs are at parity with coal, and they are unable to compete on a price basis on the open market. Given the substantial historical (and often hidden) subsidies which have gone into establishing the incumbent fossil fuel-based energy system (see Doornbosch & Knight, 2008), this presents public policy grounds for government intervention to assist in the deployment of renewable energy technologies into the market.

Although a carbon price supports demand in the market for renewable energy, renewable energy generators bear the risk of carbon price volatility. In the event that a carbon price falls too low, renewable energy generators are unable to compete with coal-fired electricity to sell into the grid. This means that early-stage investors in new renewable energy technologies may not have certainty in the end market for their electricity. A problem with this is that it may result in an under-investment in R&D into new technologies as investors are unwilling to bear the market risk involved, as well as an under-deployment of proven renewable technologies into the market. The introduction of a floor price for carbon is one way in which this risk can be mitigated by the government.

However, additional non-price fiscal policies may be used in the short run to give certainty to investors that there is demand for their new technologies in the market.

One example of a demand-side policy is the introduction of legislation which sets a minimum standard (renewable portfolio standard (RPS)) for renewable energy generation in the national energy mix. This standard sets a clear threshold for environmental performance while remaining agnostic on which technologies are used to meet this performance.

Demand-side policies have been introduced in Europe and numerous APEC countries to address market demand for clean technologies. Typically, demand-side policies are expenditure neutral on the national budget. As such, they do not fall under a strict definition of fiscal policies. Nevertheless, they are important tools in the fiscal setting of a domestic economy because they directly influence the private investment landscape within an economy and are, therefore, considered in our analysis described below.

5.1 *Expenditure-Neutral Demand-Side Measures*

5.1.1 *Targets and standards.* National targets play an important role in setting out the policy direction and aspirations of macroeconomic policy around climate change investment. They can be either mandatory, supported by legislative penalties and incentives, or merely aspirational. Although targets do not directly provide investment certainty, they provide a framework within which the private sector can be confident that the government is working towards more detailed fiscal and regulatory measures.

In the energy sector as of 2009, targets have been set in at least 73 countries globally on the basis of an agreement on the volume or proportion of the national energy mix which is to come from renewable energy sources.[3] In Table 1, a few examples of targets set by key APEC countries are presented. Targets may be set on a volume basis, a proportional basis or a technology-specific basis. For example, in Australia, it is not mandated which renewable energy technologies will deliver the outcomes set by the government.

Targets have also played a key role in the biofuel sector. Biofuels are a controversial sector because their environmental benefits are a contentious issue. Nevertheless, in terms of driving technological change, targets on the proportion of transport fuels which come from biofuels have been very effective. In most EU countries, biofuels targets are currently set at around 5.75% of transport fuels by 2010.

These specific targets give private investors certainty about the existence of a future market for biofuel products as well as help them evaluate the future size of this market. In the absence of biofuel blending targets, biofuels must compete with fossil-based fuels in the fuel market which is extremely price sensitive and where production costs price them out of the market. Table 2 presents examples of biofuel targets for a number of APEC countries.

Table 1. Examples of renewable energy targets in APEC countries

Australia	45 terawatt-hours (TWh) of electricity by 2020
Japan	14 GW of solar PV by 2020 and 53 GW by 2030
Indonesia	9.5 GW of geothermal energy by 2025
China	15% of energy from renewable sources by 2020
Vietnam	3% of commercial energy supply by 2010 and 5% by 2025 and 11% in 2050

Table 2. Examples of biofuel targets in APEC countries

Australia	350 million litres by 2010
Indonesia	3% by 2015 and 5% by 2025
Japan	500 million litres by 2012
Vietnam	300 million litres by 2020

5.1.2 *RPSs and certificates.* Another example of using domestic regulation to reduce market risk for private investors is RPSs. An RPS mandates the supply of a certain percentage of electricity to come from renewable energy generators. This obligation can be placed on different entities in the electricity supply chain, but is typically placed on electricity supply companies (e.g. in the UK, Belgium and the Netherlands). These suppliers are given a specific target on the target amount of renewable energy they must supply (10.4% target by 2010 in the UK) as well as the target power supply sources (e.g. wind power, PV, geothermal, etc.).

An RPS (or equivalent) regulation has been introduced in Australia, parts of Europe, the UK, a number of states in the US, Japan, India and Chile. It has typically been more prevalent in developed countries because of the impact on prices for the end user, but arguably has a greater role to play in APEC countries.

An RPS regulation does not in theory have a negative impact on the national budget. Rather, through regulation, it provides certainty in the market for a proportion of renewable energy to be generated and sold into the grid. Standards are usually fulfilled by issuing renewable energy generators with "credits", which can then be traded and purchased by utilities in fulfilment of the RPS mandate. This stimulates private investment because investors in renewable energy are now required to compete with each other rather than with coal-fired electricity where they have a structural price disadvantage (see Box 4).

Box 4. Examples of RPS-style standards in developing economies

India has introduced a national RES of 5% of grid power purchased by state utilities to come from renewable sources by 2010, with an increase to 15% by 2020. In total, 17 of India's 28 states have also enacted RES mandates, which vary from 1% to 20% for compliance in the period 2011–2012. One weakness in the Indian system at present is that non-compliance with RES regulations does not result in strong penalties.

In Chile, the new RPS starts at 5% for the period 2010–2014 and will increase incrementally to 10% by 2024.

In China, the national government has mandated that 15% of energy by 2020 will come from renewable energy.

As a result of being technology agnostic, an RPS policy will be advantageous to those technologies which have the lowest costs, such as wind power (Johnstone *et al.*, 2010). However, it will also facilitate competition in the market as renewable energy technologies will now be required to compete against each other on a cost basis rather than with coal-fired electricity.

5.1.3 *Feed-in tariffs.* An increasingly popular policy among policy-makers and the private sector to promote renewable energy generation has been feed-in tariffs. Feed-in

Table 3. APEC countries with feed-in tariff policies

Australia		New Zealand	
Brunei		Papua New Guinea	
Canada	✓*	Peru	
Chile		The Philippines	
China	✓	Russia	
Indonesia	✓	Singapore	
Japan	✓	Thailand	✓
Republic of Korea	✓	US	✓
Malaysia		Vietnam	
Mexico			

Note: ✓ indicates that some states and provinces in these countries have feed-in tariffs.

tariffs are popular among private investors because they offer certainty in the end market to renewable energy generators competing with coal-fired electricity to sell into the grid. Typically, feed-in tariff regulation guarantees grid access for renewable energy suppliers. In addition, it typically offers long-term contracts for electricity, which are shown in Table 3.

Although each jurisdiction has a very different mechanism for calculating the feed-in tariff rates, the cost of these policies draws down on the national budget. Some jurisdictions, such as China, differentiate the generosity of the tariff based on the technology it is seeking to promote. This is typically negotiated on a regional basis given the competitive industrial strengths of the area in question.

The disadvantage of technology-specific feed-in tariffs is that there are few spillover benefits for competing technologies. There is a risk that if this advantage persists for too long, it will have a "lock-in" effect whereby one technology becomes dominant. Short-term approaches to renewable energy feed-in tariffs have been applied in Europe. In Spain, the solar PV feed-in tariff was rapidly reduced after its 2010 targets were reached ahead of schedule. Although this created short-term pain for some companies, it has helped grow a strong solar industry in Spain (Box 5).

Box 5. Examples of feed-in tariffs in developing economies

A number of developing economies with APEC currently use feed-in tariffs. Thailand has adopted a feed-in tariff which affects particular technologies—wind, solar, biomass and micro-hydro. In Indonesia, the new revised feed-in tariff is targeted at renewable energy projects of a particular size—only plants greater than 10 MW in size are included.

5.1.4 *Consumption standards.* Finally at a local level, a government may promote the use of certain technologies through amending regulation to mandate the uptake of certain technologies. These amendments are typically to environmental and planning regulation as well as fuel efficiency standards.

For example, Spain in 2008 became the first country to mandate solar water heating at both the national and local levels. This has been followed in a number of jurisdictions including India where solar hot water is now mandated for a variety of new institutional, corporate and residential buildings. In China, the regional authorities are moving towards stricter guidelines for technology and equipment use in new buildings.

6. Other Investment Barriers to Project Finance in Developing Countries in APEC

The above discussion has covered a number of alternative approaches to carbon pricing to address investment barriers around the development and deployment of renewable energy technologies. However, when seeking to finance a renewable energy project in a developing country, project financiers and developers typically face barriers independent of the project being a clean technology. Addressing these barriers would help increase inbound foreign investment in major infrastructure projects in APEC countries.

Some of the financing challenges faced by major infrastructure projects in these developing economies include the following:

- Sovereign risk: government instability, lack of transparency in business dealings, legal enforcement. Multilateral Investment Guarantee Agency (MIGA) guarantees insuring against contract default, currency inconvertibility, expropriation and war and strife.
- Regulatory risk: change of regulatory and other fiscal conditions with the change of political administrations.
- Currency risk: exchange rate fluctuations make returns volatile and could impact investment. Governments may seek to mitigate these effects for foreign investors.
- Deal flow risk: significant information asymmetry around possible deals available, geographies and different agencies involved. Governments may introduce a public body responsible for assisting in project development and offering technical assistance.
- Operational risk: delays to project construction and approval and transaction costs associated with corruption.

A number of APEC countries have introduced policies to try to strengthen governance and address these barriers. In the Philippines, for example, the government has supported renewable energy trade missions domestically and internationally to try to attract business to the country. It has also offered investment kits to clarify the project approval process. Thailand has also recently established an investor relations office to specifically assist companies interested in operations and maintenance of energy assets (Box 6).

Box 6. Insuring against sovereign risk

The MIGA is an affiliate of the World Bank Group. It is tasked with the role of promoting foreign direct investment into developing countries. One of its tasks is to provide publicly funded insurance for inbound investment in projects exposed to high levels of political risk. Private investors are often unwilling to bear this risk when considering the finance of the project. These risks can include war and civil disturbance, expropriation, currency transfer risks and breach of contract.

In addition, renewable energy projects typically face a number of specific non-economic barriers which governments may seek to reduce in order to attract new investment. These hurdles are particularly around obstacles to grid access, poor electricity market design and administrative hurdles in the process of the Clean Development Mechanism of the Kyoto Protocol in order to secure additional revenue streams.

A number of countries within Asia have recently focused on improving their grid transmission infrastructure. For example, the Asian Development Bank is helping fund an interconnected power system between Indonesia and Malaysia. Vietnam and Cambodia are also looking to boost investment cooperation on hydropower plants between the two countries. In south Asia, discussions are underway to interconnect grids between India and Bhutan, India and Nepal and India and Sri Lanka. For renewable energy, the key challenge is finding a mechanism to accommodate decentralized producers into the power grid. Presently, decentralized producers are not permitted to feed overcapacity back to the grid and often face significant competition from conventional power plants around grid access. Easing these constraints would assist with many of the non-economic barriers to renewable energy deployment in APEC countries.

7. Conclusion

This paper has sought to lay out the basis for technology policy measures which can operate alongside carbon pricing for a well-balanced policy approach to low-carbon economic transition.

The case for non-carbon price policy approaches can be clearly made by taking an evolutionary economics perspective to clean technology investment. This emphasizes the historical embeddedness of incumbent fossil fuel technologies and the institutional barriers which impact project finance decisions. These include the under-investment of the private sector in R&D for new technologies (the spillover hurdle) and uncertainty over future market penetration and cash flow (adoption hurdle). While carbon pricing is effective at driving marginal efficiencies, it is argued here that non-carbon price-based policy is better suited at supporting early-stage technologies which are yet to find a substantial niche in the market and at moving the socio-technological system from one paradigm to another.

We seek to explain from theory and practice in the APEC region how non-price fiscal policies can be used to effectively facilitate economic transition. Two principles are worth emphasizing.

The first is that non-price fiscal policy needs to be targeted at the market failures it is seeking to address. This may vary across countries and sectors, but is typically focused on the renewable energy sector around high upfront costs to investment (supply-side barrier) and uncertainty among new renewable energy utility companies about the amount of electricity they will be able to sell into the grid (demand-side barriers). We have provided a number of examples in this paper about how governments and multilateral development banks can begin to address these problems.

The second principle governing non-price fiscal policy is that it needs to be temporary and coordinated with other policies. Often the barrier which non-price fiscal policy seeks to address is "early on" in a technology's development towards technological maturity and commercial demonstration. Once these barriers have been removed, it is important to recalibrate the policy to ensure that the utility companies and other service providers do not come to depend on unsustainable government interventions. An example of this might be the "lock-in" effect if utility companies indefinitely depend on price-related subsidies. For example, implemented before climate change was a concern, fossil fuel subsidies have become locked in and negatively distort the energy market in many countries. This leads to biases in energy investment and comes at a heavy cost to government finances. Often, the most effective policy is to remove previous policies which directly or indirectly have supported high-carbon industries. Finally, it is worth emphasizing that outside of climate policy, good governance as a crucial ingredient in fostering clean technology investment has examined some of the sovereign risk issues impeding private investment in major projects, especially within developing countries where the rule of the law may be problematic. These issues should not be underestimated because they may impact foreign direct investment across all sectors, not just renewable energy. Addressing each of these issues in strategic combination is necessary to remove the barriers to private investment and support the cost-effective transition to a low-carbon economy.

Acknowledgements

This paper has benefited significantly from the input and advice of several people, in particular I would like to thank Drs Eric Knight and Dariusz Wójcik, Professor Gordon Clark and David Imbert for their helpful input and advice – all work remains the sole responsibility of the author and the usual disclaimer applies.

Notes

1. See World Resources Institute website, http://projects.wri.org/taxonomy/term/8?page=3.
2. Further details on the structuring of publicly backed guarantees are available in the SEFI Alliance Report 'Publicly Backed Guarantees as Policy Instruments to Promote Clean Energy'.
3. See Renewable Energy World website, http://www.renewableenergyworld.com/rea/news/print/article/2009/09/renewables-global-status-report-2009-update.

References

Acemoglu, D., Aghion, P., Burstyn, L. & Hemous, D. (forthcoming) The environment and directed technical change, *American Economic Review*.

Ahmad, S. (1966) On the theory of induced innovation, *Economic Journal*, 76(302), pp. 344–357.

Arrow, K. & Debreu, G. (1954) Existence of an equilibrium for a competitive economy, *Econometrica*, 22(3), pp. 265–290.

Arthur, W. B. (1988) Self-reinforcing mechanisms in economics, in: P. Anderson, K. Arrow & D. Pines (Eds) *The Economy as an Evolving, Complex System*, pp. 9–31 (Reading, MA: Addison-Wesley).

Arthur, W. B. (1989) Competing technologies, increasing returns, and "lock-in" by historical events, *Economic Journal*, 99(394), pp. 116–131.

Arthur, W. B. (1994) *Increasing Returns and Path Dependence in the Economy* (Ann Arbor: University of Michigan Press).

Baumol, W. J. & Oates, W. E. (1988) *The Theory of Environmental Policy*, 2nd ed. (Cambridge: Cambridge University Press).

Beinhocker, E. D. (2007) *The Origins of Wealth: Evolution, Complexity and the Radical Remaking of Economics* (London: Random House).

Binswanger, H. P. (1974) A microeconomic approach to innovation, *Economic Journal*, 84(336), pp. 940–958.

Brohé, A., Eyre, N. & Howarth, N. A. A. (2009) *Carbon Markets: An International Business Guide* (London: Earthscan/James and James).

Burer, M. & Wustenhagen, R. (2009) Which renewable energy policy is a venture capitalist's best friend? Empirical evidence from a survey of international cleantech investors, *Energy Policy*, 7(12), pp. 4997–5006.

Coase, R. (1960) The problem of social cost, *Journal of Law and Economics*, 3(1) pp. 1–44.

Corden, W. M. (1997[1974]) *Trade Policy and Economic Welfare* (Oxford: Oxford University Press).

Costantini, V. & Crespi, F. (2007) Environmental regulation and the export dynamics of energy technologies, *Ecological Economics*, 66(2–3), pp. 447–460.

Cropper, M. L. & Oates, W. E. (1992) Environmental economics: A survey, *Journal of Economic Literature*, 30(2), pp. 675–740.

David, P. A. (1985) Clio and the economics of QWERTY, *American Economic Review*, 75(2), pp. 332–337.

David, P. A. (1993) Path dependence and predictability in dynamic systems with local network externalities: A paradigm for historical economics, in: D. Foray & C. Freeman (Eds) *Technology and the Wealth of Nations*, pp. 187–216 (London: Pinter).

David, P. A. (2005) Path dependence in economic processes: Implications for policy analysis in dynamical systems contexts, in: K. Dopfer (Ed.) *The Evolutionary Foundations of Economics*, pp. 151–194 (Cambridge: Cambridge University Press).

Dechezleprêtre, A., Glachant, M., Hascic, I., Johnstone, N. & Ménière, Y. (2008) *Invention and Transfer of Climate Change Mitigation Technologies on a Global Scale: A Study Drawing on Patent* Data, Working Paper, Paris: Centre for Industrial Economics.

Doornbosch, R. & Knight, E. (2008) *What Role for Public Finance in International Climate Change Mitigation*, OECD Roundtable for Sustainable Development, Paris: SG/SD/RT(2008)3.

Fankhauser, S. & Helpburn, C. (2010a) Designing carbon markets. Part I: Carbon markets in time, *Energy Policy*, 38(8), pp. 4363–4370.

Fankhauser, S. & Helpburn, C. (2010b) Designing carbon markets. Part II: Carbon markets in space, *Energy Policy*, 38(8), pp. 4381–4387.

Fischer, C. (2009) The Role of technology policies in climate mitigation, *Issues Brief*, Resources for the Future, Washington, DC.

Foxton, T. (2011) The rationale for policy interventions from and innovations systems perspective, in: J. Murphy (Ed.) *Governing Technology for Sustainability,* Ch. 7, pp. 129–147 (London: Earthscan).

Geels, F. (2002) Technological transitions as evolutionary reconfiguration processes: A multi-level perspective and a case-study, *Research Policy*, 31(2002), pp. 1257–1274.

Geels, F. W. (2005) *Technological Transitions and System Innovations: A Co-evolutionary and Socio-Technical Analysis* (Cheltenham: Edward Elgar).

Geels, F. & Kemp, R. (2006) Transitions, transformations, and reproductions: Dynamics in socio-technical systems, in: M. McKelvery & H. Holmén (Eds) *Flexibility and Stability in the Innovating Economy*, pp. 227–256 (Oxford: Oxford University Press).

Grubb, M. & Ulph, D. (2002) Energy, the environment and innovation, *Oxford Review of Economic Policy*, 18(1), pp. 92–106.

Hazilla, M. & Kopp, R. (1990) Social cost of environmental regulations: A general equilibrium analysis, *Journal of Political Economy*, 98(4), pp. 853–873.

Helm, D. (2005) Economic instruments and environmental policy, *The Economic and Social Review*, 36(3), pp. 205–228.

Hepburn, C. (2006) Regulation by prices, quantities or both: A review of investment choice, *Oxford Review of Economic Policy*, 122(2), pp. 226–247.

Horbach, J. (2007) Determinants of environmental innovation: New evidence from German panel data sources, *Research Policy*, 37(10), pp. 163–173.

Indonesian Green Paper (2009) *Ministry of Finance Green Paper: Economic and Fiscal Policy Strategies for Climate Change Mitigation in Indonesia* (Jakarta: Republic of Indonesia).

Jacoby, H. D. & Ellerman, A. D. (2004) The safety valve and climate policy, *Energy Policy*, 32(4), pp. 481–491.

Johnstone, N., Hascic, I. & Popp, D. (2010) Renewable energy policies and technological innovation: Evidence based on patent counts, *Environmental Resource Economics*, 45(1), pp. 133–155.

Jorgenson, D. W. & Wilcoxen, P. J. (1990) Environmental regulation and US economic growth, *RAND Journal of Economics*, 21(2), pp. 314–340.

Kamien, M. I. & Schwartz, N. L. (1968) Optimal induced technical change, *Econometrica*, 36(1), pp. 1–17.

Kemp, R. (1994) Technology and the transition to environmental sustainability. The problem of technological regime shifts, *Futures*, 26(10), pp. 1023–1046.

Kemp, R., Rip, A. & Schot, J. (2001) Constructing transition paths through the management of niches, in: R. Garud & P. Karnoe (Eds) *Path Dependence and Creation*, pp. 269–299 (Mahwah, NJ: Lawrence Erlbaum).

Kemp, R., Schot, J. & Hoogma, R. (1998) Regime shifts to sustainability through processes of niche formation: The approach of strategic niche management, *Technology Analysis and Strategic Management*, 10(2), pp. 175–196.

Knight, E. R. W. (2010) *The Economic Geography of Clean Tech Venture Capital*, Working Paper (available on SSRN). Oxford: Centre for Employment, Work and Finance, School of Geography and Environment, Oxford University.

Kriechel, B. & Ziesemer, T. (2009) The environmental porter hypothesis: Theory, evidence, and a model of timing of adoption, *Economics of Innovation and New Technology*, 18(3), pp. 267–294.

Lanoie, P., Patry, M. & Lajeunesse, R. (2008) Environmental regulation and productivity: Testing the porter hypothesis, *Journal of Productivity Analysis*, 30(2), pp. 121–128.

Levinthal, D. A. (1998) The slow pace of rapid technological change: Gradualism and punctuation in technological change, *Industrial and Corporate Change*, 7(2), pp. 217–247.

Martin, R. (2010) Roepke lecture in economic geography—rethinking regional path dependence: Beyond lock-in to evolution, *Economic Geography*, 86(1), pp. 1–27.

Ministry of Finance Green Paper (2009) *Economic and Fiscal Policy Strategies for Climate Change Mitigation in Indonesia*.

Nelson, R. R. & Winter, S. G. (1982) *An Evolutionary Theory of Economic Change* (Cambridge, MA: Belknap Press).

Newell, R., Jaffe, A. & Stavins, R. (1999) The induced innovation hypothesis and energy-saving technological change, *The Quarterly Journal of Economics*, 114(3), pp. 941–975.

Nordhaus, W. (2007) To tax or not to tax: Alternative approaches to slowing global warming, *Review of Environmental Economics and Policy*, 1(1), pp. 26–44.

Olson, M. (1965) *The Logic of Collective Action* (Cambridge, MA: Harvard University Press).

Palmer, K., Oates, W. E. & Portney, P. R. (1995) Tightening environmental standards: The benefit cost or the no-cost paradigm, *Journal of Economic Perspectives*, 9(4), pp. 119–132.

Philibert, C. (2009) Assessing the value of price caps and floors, *Climate Policy*, 9(6), pp. 612–633.

Pigou, A. C. (1912) *Wealth and Welfare* (New York: Macmillan).

Pigou, A. (1932) *The Economics of Welfare*, 4th ed. (London: Macmillan).

Popp, D. (2006) Comparison of climate policies in the ENTICE-BR model, *The Energy Journal*, Special Issue: Endogenous Technological Change and the Economics of Atmospheric Stablisation, 1, pp. 163–174.

Popp, D. (2010) The Porter hypothesis at 20: Innovation and environmental regulation, presentation given to the Maxwell School of Citizenship and Public Affairs, Syracuse University.

Porter, M. E. & van der Linde, C. (1995) Toward a new conception of the environment-competitiveness relationship, *Journal of Economic Perspectives*, 9(4), pp. 97–118.

Rip, A. & Kemp, R. (1998) Technological change, in: S. Rayner & E. L. Malone (Eds) *Human Choice and Climate Change,* (vol. 2, pp. 327–399 (Columbus, OH: Battelle Press).

Schot, J. & Geels, F. W. (2008) Strategic niche management and sustainable innovation journeys: Theory, findings, research agenda, and policy, *Technology Analysis and Strategic Management*, 20(5), pp. 537–554.

Schot, J. W., Hoogma, R. & Elzen, B. (1994) Strategies for shifting technological systems. The case of the automobile system, *Futures*, 26(10), pp. 1060–1076.

Schumpeter, J. (1975[1942]) *Capitalism, Socialism and Democracy* (New York: Harper and Row, Harper Colophon Ed).

Shellenberger, M., Nordhaus, T., Navin, J., Norris, T. & Van Noppen, A. (2008) Fast, clean, and cheap: Cutting global warming's Gordian Knot, *Harvard Law & Policy Review*, 2(1), pp. 93–118.

Simon, H. A. (1955) A behavioral model of rational choice, *Quarterly Journal of Economics*, 69(1), pp. 1–18.

Simon, H. A. (1957) *Administrative Behavior: A Study of Decision-making Processes in Administrative Organization* (New York: MacMillan).

Stern, N. (2006) *The Economics of Climate Change – The Stern Review* (Cambridge: Cambridge University Press).

Stern, N. (2009) *A Blueprint for a Safer Planet: How We can Save the World and Create Prosperity* (London: Vintage).

Unruh, G. C. (2000) Understanding carbon lock in, *Energy Policy*, 28(12), pp. 817–830.

Yang, C.-J. & Oppenheimer, M. (2007) A "Manhattan Project" for climate change? *Climate Change*, 80(3–4), pp. 199–204.

Index

Note:
Figures and illustrations are in italics
Tables are in bold

1G biofuels 53
2G biofuels 53, 59
2005 Energy Policy Act (US) 154
2006 Stern Review on the economics of Climate Change 144, 145
2009 American Recovery and Reinvestment Act 154
2009 Copenhagen Conference on Climate Change 1

accelerated depreciation 153
actors in technological regimes 66
AC wind turbines 37–8, 39
"adjacent possible" and innovation 91, 92, 98, 102, 103
"adoption hurdle," the, as investment barrier 150, 151, 155, 160
APEC (Asia Pacific Economic Cooperation) region 144, 151, 158, **158,** 159–60; energy targets in **156,** 156–7, **157**
ATV (Academy of Technical Sciences) 13
Auken, Svend 19–20

biofuels 156, **157**; and path creation **52,** 52–8, **54–5,** 59, 100–2
biogas energy 98
biomedical applications of products 101
blocking mechanisms in TIS 47
Bonus Wind Power 15, 19
breaks with technologies and products 113
"bricolage," idea of 33

Californian wind power industry, the 16–17, 19, 20, 99 *see also* wind turbine manufacturers in the United States
Californian wind turbine design 99
capitalism as economic change 147–8
carbon lock-in and new energy technologies 59–60
carbon pricing: and clean energy technologies 144, 146–7, 149–1503–145, 150; in renewable energy market 151, 155
Carbon Trust, the 83
change in Danish wind power policy in 1990s 19–20
Chinese tax incentives for energy projects 153
clean energy technologies and carbon pricing 144, 146–7, 149–50 *see also* RETS (renewable energy technologies)
climate policy 144, 146–7, *149,* 149–50
clusters and cluster research 6 *see also* Danish wind turbine cluster, the; regional clusters in RETS
Coasian approach and pollution, the 144
cognitive barriers to new path creation 30–1
comparison between energy technology path creations 50–1
conversion of a technology 34, 42
creative processes in new path creation 91
critical mass and turning away from path creation system 35
currency risk 159

Danish shift to decentralized CHP (combined heat and power) stations 96–100, *97*
Danish wind turbine cluster, the 5–6, 7–12, **9,** *11,* **12,** 20–2, 76 *see also* North Jutland's green innovation system
David model of 'QWERTYnomics' 26–7
deal flow risk 159
DECC (Department of Energy and Climate Change) 82, 83
decentralization of the energy system 67
DEF (Association of Danish Electric Utilities) 11, 12
de-locking mechanisms for industries 115–16
demand factors influencing the location of wind turbine companies 131–3
demand-side measures to support energy 155–9, **156, 157, 158**
Denmark and transitional processes in renewable energy *see* empirical analysis of renewable energy policy
deployment in renewable energy 71, 72, *73,* 75, 82

deployment of RETs in Navarra 78–80, *80,* 85
design driven innovation 94
design-driven innovation 103
diffusion and innovation 35, 38–9, 42
discourse articulation and social change 98–9
displacement 33
"diversity," Jacobian concept of 90
Domsjö 101, 102
drivers in growth of German offshore wind energy sector 119
Dutch banking 94
DWO (Danish Windmill Owners Association) 11

eco-innovation 97, 98, *98*
economic analysis of climate change 144, 146–7, *149*
economic dimension and energy innovation **54–5,** 55–6, 60
economics of wind power, the 17–18
EEG (Renewable Energy Sources) Act 119
EHN (Energia Hidroelectrica the Navarre) 79
electricity generation in the UK 66
electricity grid and wind turbines, the 11, 37, 38, 40
electricity prices and wind power generation 131, 139
electricity supply from renewable energy 157
electricity supply in Denmark 39–41, *40, 41,* 43
empirical analysis of energy technology path creations **48, 49,** 50–3, **52,** 56–60; differences in SNM and TIS biofuel analyses 53–6, **54–5**
empirical analysis of renewable energy policy 71–6, *73,* **74,** *75*; potential for regional niche development 77–8, *78,* 80 *see also* UK renewable energy policy
energy policy in Denmark 12–15, 37, 39, 76; adoption of decentralized CHP stations 96–100, *97 see also* Danish wind turbine cluster, the; North Jutland's green innovation system
energy production and innovation 45, 72
energy targets **156,** 156–7
entrepreneurs: and innovation 47, 90, 95–6; and new path creation 32, **48**
environmental policies and innovation 146
environmental technology industries *see* location preferences of new industries
ergodicity and path dependence theory 27, 111
EU directives on biofuels 57
EU energy policy 43
European Investment Bank, the 154
evolutionary economic geography 92–3
evolutionary innovation economics 90, 92, 147, *149,* 160
evolutionary urban economics 90 *see also* Jacobs, Jane
expenditure-neutral demand-side energy measures **156,** 156–9, **157, 158**
export of Danish wind turbines 16

financial investment in German offshore wind energy 119
Finland and biofuel **54–5,** 57
FIT (feed-in-tariffs) system 66, 76, 157–8, **158**
funding of regions for renewable energy 80–2

Geothermal Energy Development Fund, the 154
geothermal resources in Indonesia 152
Germany and transitional processes in renewable energy *see* empirical analysis of renewable energy policy
global carbon emissions 63–4
government interventions of energy pricing 145, 146, 147, 151, 155–6, 160
government support for a new industry 132–3, 139, 152, 153–4, 158–9
grid transmission infrastructure in Asia 160
Grove-Nielsen 10, 15, 16, 21

Hamburg and the German offshore wind energy industry 121, 122
Hayek, Friedrich 145
Herborg Wind Power 15
Herborg wind turbine design 10
heterogeneity as a de-locking mechanism 115
high start-up costs of renewable energy technologies 152–3
human capital for the German wind energy industry 116–17
hybrid forums and creation-type processes 22
hybrid theory 32–4, 35–6, *36,* 41

Indonesian renewable energy policy 152, 155
induced innovation 145–6
infrastructure for German wind energy industry 116–17
innovation 94–6, 146; and the "adjacent possible" 91, 92, 98, 102, 103; and energy production 45, 72; and entrepreneurs 47, 90, 95–6; incremental innovation 99, 100–2; and new path creation 33, 35, 90–1; niche innovations **49,** 49–50, 66, 69
institutional churn and wind turbine technology 39
institutional hysteresis 31
inventions in niche environments 34–5
inventors and R&D and new path creation 32–3, 42

investment barriers in clean energy technologies 160–1; from the demand side 155–9, **156, 157, 158**; from the supply side 150–5, 159, 160
investment tax credits 153

Jacobs, Jane 90, 94, 96, 102, 103
Jacobs' externalities 94
Juul, Johannes 37; wind turbine design 8, 10, 21

KK Electronics 15

la Cour, Poul 37, 38, 40
landscapes: affect on regimes 69–70, 71; and change in electricity supply in Denmark 39–41, *40, 41,* 43
layering to an existing technology 34, 40, 42
LM Fiber Glass 16
loan schemes for energy projects 153–4
location preferences of new industries 112, 129–33; of US wind turbine manufacturers 132, *133,* 133–40, **134–5, 136,** *136,* **137–8**
lock-in of sub-optimal technologies as industry standards 27, *27,* 30, 158
Lombardy regional innovation system 95
Lykkegaard Company, the 38

marine energy technology 83
market interventions and energy prices 145
market risks in financing of renewable energy projects 151–2
market uncertainty in clean technologies market 155, 160
Meyer, Niels 13, 18
MIGA (Multilateral Investment Guarantee Agency) 159
"mindful deviation" from new creation paths 28, 34–5, 37–8, 40
MLP (multi-level perspective) 45, 46, 49, 64, 148–9; as energy transition process 65–7, 69–71, *71,* 74 *see also* SNM/MLP (strategic niche management/multi-level perspective)
Morrás, Esteban 79
motor car, development of the 33
motors of innovation 47

Navarra and renewable energy clusters 78–80, *80,* 85
neo-Schumpeterian theories on evolution of economies 25–6
Netherlands, the, and biofuel 53–5, **54–5,** 56–7
Neuer Hafen Kiel-Canal, the 111
new industrial development paths, origin of 112–15, **115** *see also* offshore wind energy in Northern Germany
"new orthodoxy" and innovation 103
new path creation: and innovation 33, 35, 90–1; technological barriers to 31, 34, 35–6, *36,* 39, *41, 42*; theoretical possibilities of 1–2, 91 *see also* regional path creation
niche development in RETS *77,* 77–8
niche innovations **49,** 49–50, 66, 69
niches by region 70–1, *71*
niches for new technologies 34–5, 37, 38, 48–9
NIVI (North Jutland Institute for Renewable Energy) 10
non-carbon pricing policy 160
non-ergodicity 27, 112
Nordtank 15
Norrland, Sweden and incremental innovation 100–2, *101*
North Jutland's green innovation system *98,* 99–100, 103 *see also* Danish wind turbine cluster, the
nuclear power 12, 13, 99

offshore wind energy in Northern Germany 107–8, 110–12 *110,* 115, 116–23
OOA (Organization for Information about Nuclear Power) 12, 17
operational risk 159

patent data 72, *73,* 74, **74,** 75; renewable energy patents *77,* 77–8, **78,** 80
path creation processes 7–8, 20–1, 28–9, *29,* 32–6, *36,* 41–2; and biofuels **52,** 52–8, **54–5,** 59; and wind power technologies in Denmark 37–40, *40,* **41,** 42 *see also* regional path creation
path dependence theory 6–7, 26–32, *27,* 42, 68, 89–90, 148; and dominant regimes 66; and ergodicity 27, 111; and innovation 91–3, *92, 93,* 103; by region 96–100, *97, 98*
path-dependent barriers to new path creation 32, 36, *36,* 39, 42, 43
path inter-dependence, notion of 93
'peak cotton,' era of 101
Pedersen, Helge 10, 13
Peterson, Amdi 38
Pigouvian tax, the 144
pollution and carbon pricing 144
"preadaptation" 91, 92, 97, 102
process theory 51
Processum biorefinery hub 100–2, *101*
production costs of renewable energy generation 154–5
proximity to demand as a spatial pattern 131
public sector bank guarantees for energy infrastructure 154

railroads, the, and innovation 95

R&D specialisation in the German wind energy industry 121
R&D trends in renewable energy 71–2, *73,* 74–6, 84, 152
regional clusters in RETs 64–5, 78, 80; in North Jutland *98,* 99–100; in the UK 81–4, *82*
regional evolution and "related variety" 90, 92, 93, 95, 96
regional innovation 92, *93,* 96, 99–100
regionalization of national energy regimes 67–8, 70
regional path creation 68–9
regional path dependence 96–100, *97, 98*
regulations on wind power 17–19, 22
regulatory landscape for RETs *75,* 75–8
regulatory risk 159
"related variety," and regional evolution 90, 92, 93, 95, 96
renewable portfolio standards 139
report on wind power in Denmark 13
reproduction as change at socio-technical regime level 148–9
resource allocation in a perfectly competitive market 147
restrictions on investment and ownership in wind turbines 18
RETS (renewable energy technologies) 1, 3, 64, 71, *75,* 75–6; deployment of in Navarra 78–80, *80,* 85; niche development in *77,* 77–8 *see also* clean energy technologies and carbon pricing
Riisager Wind Power 15–16
Riisager wind turbine design 8–9, 11, 21
"routine" and business processes 148
RPS (renewable portfolio standard) 156, 157
RTA (revealed technological advantage) 74

Schumpeter, Joseph, and innovation 95
shipbuilding in northern Germany 108, 109–10, *110,* 116–17
shipyard crisis in northern Germany 114, 116; diversification into wind energy 120–1, 122
slump in the Danish wind turbine market 19
SNM/MLP (strategic niche management/multi-level perspective) 45, 48–9, 50
socio-technical change 148–9
socio-technical regimes 65, 66, 67
solar home systems, financing of 154
solar power energy in the U.S. 153, 154
solar water heating in Spain 158–9
sourcing of turbine components 16
sovereign risk 159, 161
Spain and transitional processes in renewable energy *see* empirical analysis of renewable energy policy
spatial development of technological pathways 2–3
spatial dimension, the, and energy innovation 56–8, 60
"spillover hurdle," the 150, 160
state intervention in wind power 5–6
Stern, Nicholas 144, 145
Stiesdal, Henrik 10, 15, 39
subsidy-approval scheme for windpower in Denmark 14–15, 17, 99
supplier competencies for wind turbines 15–16, 131–2, 139
supply-side measures to support technology 152–5
Sweden and biofuel 53, **54–5,** 55, 56–7, 59, 100–2
Swedish innovation strategy 101

tax holidays 153
technological barriers to new path creation 31, 34, 35–6, *36,* 39, *41,* 42
technological paradigms and new path creation 30–2
technological regimes 31–2, 65
"technology push" innovation 95
TISs (technological innovation systems) 45, 46–8, **48**
tractor technology 91
transformation as socio-technical change 149
transition as a socio-technical change 149
transition research in energy 69, 70–1, *71,* 74, 84
translation of wind power 21
transplantation as a de-locking mechanism 115–16
"transversality" 92, 93, 96, 97
TRC (test station for wind power) 13–14
Tvind wind turbine design 10, 37–8

UK renewable energy policy 64, 65, 68, 70, 84; empirical analysis of 71–7, *73,* **74,** *75;* potential for regional niche development *77,* 77–8, 80; by region 81–4, *82*
user-driven innovation 94

Vestas 16, 21, 38–9, 76, 79, 97–8

wave and tidal technology 83
Wave Hub 82–3
welfare economics in a perfectly competitive market 147
wind energy policy of Bremen 117–18
wind power in Navarra *see* deployment of RETs in Navarra
wind power technologies in Denmark 37–40, *40,* **41,** 42; regulations on 17–19, 22;

turbines and the electricity grid 11, 37, 38, 40
wind resources in the US 131
wind turbine designs in Denmark 8–11, **9,** *11,* 13–15, 17, 19, 20, 99; early innovations in 37–9; and supplier competencies 15–16, 131–2, 139; Tvind design 10, 37–8
wind turbine manufacturers in the United States 132, *133,* 133–40, **134–5, 136,** *136,* **137–8**; Californian and wind power 16–17, 99

yield rates for the same technology 151

FSC
www.fsc.org
MIX
Papier | Fördert gute Waldnutzung
FSC® C083411

Kolibri 360 GmbH
Ferdinand-Jühlke-Straße 7
99095 Erfurt, Deutschland
produktsicherheit@kolibri360.de